POLISONICS

Collective Listening and the Politics of Sound

Edited by Matt Lewis and Manal Massalha

Routledge
Taylor & Francis Group
LONDON AND NEW YORK

Designed cover image: Cecilia Bonilla

First published 2026
by Routledge
4 Park Square, Milton Park, Abingdon, Oxon OX14 4RN

and by Routledge
605 Third Avenue, New York, NY 10158

Routledge is an imprint of the Taylor & Francis Group, an informa business

British Library Cataloguing-in-Publication Data
A catalogue record for this book is available from the British Library

ISBN: 978-1-032-04734-8 (hbk)
ISBN: 978-1-032-04732-4 (pbk)
ISBN: 978-1-003-19445-3 (ebk)

DOI: 10.4324/9781003194453

Typeset in Sabon
by SPi Technologies India Pvt Ltd (Straive)

Access the Support Material: www.routledge.com/9781032047324

POLISONICS

Bringing together contributors with backgrounds in ethnography, theatre, design, music, and sound, *Polisonics: Collective Listening and the Politics of Sound* demonstrates the importance of sound in design and how the sonic offers alternatives to dominant, singular, and technologically determined narratives.

From sonic street interventions to rural field recording, water fountains to community listening sessions, and Palestinian Hip-Hop to sound for VR, this book foregrounds participatory modes of making and collective engagement, particularly those rooted in musical cultures. It examines how relational approaches to sound can help reshape paradigms and challenge systemic inequalities, build communities, and resist structural injustice.

With chapters on areas including audio description, sound systems, and the use of machine listening as a tool for immigration control, *Polisonics* will be of interest to activists, academics, and musicians as well as those studying and working in sound design, ethnography, social justice, and theatre.

Matt Lewis is an artist and researcher based in Margate and London. Informed by a background in music, he is passionate about the importance of coming together in groups to make things happen. He has a PhD from Goldsmiths and works at the Royal College of Art in London, where he teaches sound and is part of the Voicing Sound research group.

Manal Massalha is a London-based sociologist, urban ethnographer, and social documentary photographer/videographer. Manal holds a PhD from the London School of Economics, and her practice explores the everyday and the mundane, the human and the urban, the cultural and the social, and the political and the economic.

CONTENTS

ABOUT THE AUTHORS

Rosie Ash is a graduate from the Royal College of Art, and now works in a school in East London as a Higher Education Co-ordinator, supporting students to access university-level education and training. Rosie works with Compass Collective as a welfare support volunteer. Previously, Rosie has worked with various migrant organisations across London, including the Say It Loud Club and the British Red Cross, as a music teacher, sound designer, and facilitator to create projects celebrating and empowering migrant voices to speak up against political rhetoric that dehumanises people seeking asylum in the UK. Rosie brings her passion and dedication to the arts to all her work, with the goal of encouraging all young people to explore and access their creative talents regardless of their background.

Harshadha Balasubramanian has a PhD from the Centre for Digital Anthropology at UCL and is an Associate Lecturer at the Royal College of Art. Harshadha's PhD, funded by the AHRC through the London Arts and Humanities Partnership, ethnographically explored the experiences of artists designing non-visual access to three-dimensional virtual environments. As a blind researcher examining visual details, Harshadha invites her research collaborators to describe everything they want her to see, including their practices, settings, and design outcomes. Harshadha also designs against the normalisation of disability lived experience, such as through interning with Microsoft Research to produce design guidelines for non-visual access to virtual environments. She consults for artists and curators, helping them to embrace the creative, multisensory possibilities unlocked by accessible design while meeting any specific access requirements within disability communities.

Mariana Carvalho (São Paulo/Berlin) is a performer, sound artist, and musician working with the body and interactions between bodies, mostly through voice, technological mediation, improvisation, text, internal listening, contact, materials, self-built instruments, and prepared piano. Carvalho holds a master's degree in Sound Studies and Sonic Arts – UdK Berlin and a bachelor's in piano at the University of São Paulo, where she was part of Sonora: músicas e feminismos and NuSom – Research Center on Sonology of USP. Collaboration is a big part of her work, most lately happening in the fields of performative arts and dance.

Brian D'Aquino is a researcher, DJ, and music producer based in Naples, Southern Italy. He is the author of *Black Noise: Tecnologie della Diaspora Sonora* (Meltemi 2021) and co-editor of the *Mobile Music Machines* book series (forthcoming, Bloomsbury). He holds a PhD in International Studies from the University of Naples L'Orientale. Between 2021 and 2025, he worked as a Senior Research Assistant on the ERC-funded project *Sonic Street Technologies* (SST) at Goldsmiths, University of London, carrying out extensive fieldwork in the UK, Jamaica, and Colombia. His research engages with Black music and sound system cultures, with a focus on practice-based epistemologies, community empowerment, bottom-up technological innovation, and the politics of noise in urban space. Since 2004, he has run the Bababoom Hi-Fi sound system, performing widely across Europe and beyond, and releasing vinyl records on the Bababoom Hi-Fi imprint.

Nathanael Kliebhan (he/him) is a Togolese-German sound artist and interdisciplinary researcher whose work explores how sound can be used to unsettle and reimagine colonial spatial configurations. Drawing on influences from jazz and electronic music, he experiments with composition, narrative, and improvisation, using audio technologies to amplify personal and collective experiences and to form tentative spaces of transcendence. Currently based in London, he studied Music at Goldsmiths, University of London, and holds an MA in Sound Design from the Royal College of Art, where he was awarded the Sir Frank Bowling Scholarship. He is completing an MSc in History and Philosophy of Science at University College London, where he is a recipient of the Aristotle Scholarship. His work has been performed at venues including Centre Pompidou in Paris, Somerset House, South London Gallery, and Café OTO.

Matt Lewis is an artist and researcher based in Margate and London. Informed by a background in music, he is passionate about the importance of coming together in groups to make things happen. Matt has collaborated with dancers and choreographers, written music for film and theatre, and produced a wide variety of sound installations. His work has been shown at

Turner Contemporary, Ars Electronica, Whitechapel Gallery, Café Oto, The Roundhouse, VOID Gallery, IRCAM, Diapason NYC, and MK Gallery. He has a PhD from Goldsmiths and works at the Royal College of Art in London, where he teaches sound and is part of the Voicing Sound research group.

Manal Massalha is a London-based sociologist, urban ethnographer, and social documentary photographer/videographer. She holds a PhD from the London School of Economics and is a Lecturer at Goldsmiths. Manal's practice explores the everyday and the mundane, the human and the urban, the cultural and the social, and the political and the economic. Manal's work has appeared in *The Guardian*, *Le Monde Diplomatique*, *The Observer*, *UN publications*, *e-flux*, *Qalandiya International (Palestine Art Biennale), Autograph, LSE photography exhibitions, and Hackney Museum*. A selection of images is part of the Museum of London collection. She has also worked for United Nations agencies in Palestine and Nigeria.

Laura Mello is a Berlin-based composer, performer, and interdisciplinary artist. Her work emerges from listening to the in-betweens, using sound to establish unheard connections. She explores the auditory aspects of language to deconstruct and reconstruct (de-)colonial perspectives, challenging dominant epistemologies. Her projects take diverse forms and often grow from interdisciplinary, site-specific research. As part of the Berlin collective Errant Sound, she curates projects like the Dystopia Sound Art Biennial. After studying Advertising, Composition, Conducting, and Musical Aesthetics in Brazil, she pursued Electroacoustic Composition in Vienna. In Berlin, she pursued an interdisciplinary PhD on the software EyesWeb and its role in interactive performances with sound, image, and movement. She received a working grant for sound art and new music from the Berlin Senate Department for Culture and Europe.

Shintaro Miyazaki (b. 1980, Prof. Dr.) teaches and researches in the field of media studies (digitality, computation, critique, and social transformation) at Humboldt-Universität zu Berlin. Until 2020, he was a senior researcher at the Critical Media Lab, FHNW Academy of Art and Design in Basel, Switzerland. He was born in former West Berlin and grew up in Basel, Switzerland.

Pedro JS Vieira de Oliveira is a sound artist and researcher whose work is committed to a study of listening and its material intersections with the violences of the European border. His work was exhibited and performed at the Akademie der Künste Berlin, Send/Receive Festival Winnipeg, CTM Festival, Festival Novas Frequências, Museo del Mare Palermo, among others. He held fellowships at the Leuphana Institute for Advanced Studies, the Helsinki Collegium for Advanced Studies, and taught at the Humboldt University of Berlin and

Heinrich-Heine University Düsseldorf. Currently, he is a guest faculty in Sound Studies and Sonic Arts at the Universität der Künste Berlin. He holds a PhD from the Universität der Künste Berlin.

Nirah Pomar is a Brazilian multidisciplinary artist, mother, singer, songwriter, music therapist, and researcher of the possibilities and potential of the voice – especially women's voices – as a means of breaking through oppression and overcoming trauma. She worked as a vocal coach for the group Vozes da Útera and took part in the Mulher Artista Resiste (Women Artists Resist) immersion, part of the Sesc Cultura ConVIDA project. She holds a PhD in Music from the State University of Santa Catarina, with research focused on musicographic archives of centenary bands, and is also engaged in the preservation of music as cultural heritage.

Elif Gülin Soğuksu is an electroacoustic composer, performer, and researcher born in Istanbul and currently based in Berlin. She recently obtained her master's degree at the Institute of Sonology in The Hague (NL). Elif draws inspiration from the fugitive behaviour and cross-modality of imagination in sound; her compositional practice grounds on catching the sounds of the mind. She researches essentially on the human voice as a potent force in the formation of mentally produced sounds through the polyphony of theoretical and practical frameworks. Her electroacoustic pieces and live performances bridge algorithmic approaches, including voice-driven live electronics and creative coding, digital processing, acousmatic music, and spatial sound.

Cecilia Wee (she/they) is a social practice artist, independent curator, researcher, educator, equality activist, and union organiser who grew up in Thatcher's London. Their research-based creative practice aims to create experimental strategies for learning, resistance, and flourishing with collaborators from historically marginalised backgrounds. Cecilia has curated exhibitions, edited books, and led research with organisations including Arts Council England, NHS Grenfell Health and Wellbeing Service, Resonance FM, and Tate Modern. She is part of a collaborative duo (CWxWC) with artist and activist William Crosby, investigating the politics of listening within cultural, social, economic, and environmental crises. Cecilia is a Tutor (Research) at the School of Communication, Royal College of Art.

ACKNOWLEDGEMENTS

It would have been difficult to put this book together without the following people, whom we wish to thank for their generosity, guidance, and inspiration: Joe Hirst, Mijke van der Drift, Mark Braniff, and Thomas Cridford.

Big thanks also to the RCA UCU Committee for their solidarity, listening, and noise-bringing skills. Thanks also to the staff and students at the RCA, particularly participants in the Polisonics Listening Sessions down in the sound studio. Thanks also to SomaRumor and, in particular, Giuliano Obici.

Finally, thanks to the support of our families for their love and patience.

INTRODUCTION

This is not another book about sound and its materiality in the abstract realm. Nor is it a collection of musings on listening phenomenology, detached from the socio-political and environmental forces that shape the sonic. It is, however, a book that brings together a collection of voices, both of writers and their collaborators, to celebrate the critical potential of the sonic as a connective, urgent, and contingent agent.

In the following chapters, we set up a series of feedback loops between different forms of sonic knowledge in the fields of art, design, ethnography, photography, theatre, and music. With examples from sites including London, Palestine, Istanbul, Berlin, and Florianopolis, we explore how a vibrational continuum[1] might offer possibilities of bringing issues and communities together in different ways. And how the sonic might give us a deeper understanding of pressing issues around systemic class and social, racial, and environmental inequalities.

In a time when coloniality and oppression persist, the genocide of Palestinians is condoned, abated, and enabled, the international rule of law is at risk, and the binary discourse of us vs. them is dominant in mainstream media. Sound, as explored below, becomes a form of reconnection, reappropriation, and a means of building solidarity creatively. The omnipresent noises emitted by Israeli Zanana[2] drones throughout its genocidal campaign in Gaza, hovering over displaced Palestinians, have been one example of sonic terror that, thanks to citizen reporting on social media, have become global vibrations, helping people from around the world to connect with the Palestinian experience.

Initiatives such as Radio Alhara, a Palestinian, Bethlehem-based independent online radio station operating since 2020, with DJs contributing

DOI: 10.4324/9781003194453-1

worldwide, continue the tradition of sonic assertion as producing an anti-colonial space. Examples in mind include Radio Free Belfast and Radio Free Derry[3] from Ireland of the 1960s and 70s. The Voice of Free and Combatant Algeria in the 1950s[4] and Radio Alice in Italy from 1976 to 1977.[5] The acoustics of these initiatives and collectives are not ephemeral, nor do they disappear when broadcasting stops. They are visceral and permeative. Their vibrational acts go way beyond the ear. They metamorphose into social and political transformation.

Drawing on many overlapping disciplines, including music, theatre, philosophy, design and communication, acoustics and fine art, sound studies has, over the last 50 years, emerged as a separate field in itself. The dominant knowledge produced, however, privileges a Eurocentric, compartmentalised understanding of the sonic that is largely rational and overlooks interdependencies and the intersectionality of nature and the environment. It therefore (re)produces hierarchies of listening which mirror class and racial bias found in much European and American literature. Key theories, such as Pierre Schaffer's [6] modes of listening, apply a visual and prescriptive framework through which to understand our listening experience. These include guided stages (modes) on how to listen, which rely on an "auraltypical" (Drever, 2022) approach to listening. Murray Schaffer's[7] work on soundscape studies (Schaeffer 1977) promotes the role of sound in environmentalism and the need to protect the natural world, but as noted by Goodman (2010), by categorising sounds into "low-fi" and "hi-fi", Schaffer, however, privileges some sounds over others, thereby setting up binary oppositions. Nature vs. man-made noise. High culture vs. low.

In recent years, however, there has been a growing focus on understanding and theorising sound from a critical perspective that questions hierarchical understandings of listening and the sonic. One that engages with embodied, horizontal experiences, social and racial justice, politics of representation, and the potentiality of the sonic to be responsive to the social and political. Notable literature includes *Sonic Bodies* (Julian Henriques, 2010); *Sonic Possible Worlds: Hearing the Continuum of Sound* (Salome Voegelin, 2017); *The Political Possibility of Sound* (Salome Voegelin, 2018); *Sonic Agency: Sound and Emergent Forms of Resistance* (Brandon LaBelle, 2018); *Acoustic Justice: Listening, Performativity, and the Work of Reorientation* (Brandon LaBelle, 2020); *The Race of Sound: Listening, Timbre, and Vocality in African American Music* (Nina Sun Eidsheim, 2018); *Sensing Sound* (Nina Sun Eidsheim, 2015); *Making It Heard: A History of Brazilian Sound Art* (Chaves & Lazzetta, 2019); *Sound Practices in the Global Sout*h (Buhaditya Chattopadhyay, 2022); *Aural Diversity* (John Drever, 2022); *Deaf Jam* (Mara Mills, 2008); *Testing Hearing: The Making of Modern Aurality* (Mara Mills); *Diminished Faculties: A Political Phenomenology of Impairment* (Jonathan Sterne, 2024); *Sonic Rebellions: Sound and Social Justice* (Canton,

2024); and *Alto e bom som: uma história da arte sonora brasileira.* (Chaves & Lazzetta, 2022)

The collection of chapters in this book, therefore, aims to add to this growing body of critical literature. Breaking away from reductionist approaches to sound studies, we explore themes where sound is multisensory, not limited to the ear. Sound as operating within an ecology of different environments, including the bodily, participatory, and community-centred design practices, media, material, and non-material, which intersect and converge, and connect us to the political, social, cultural, counter-cultural, and colonial.

The order of the chapters in this book is not hierarchical in importance, nor linear in narrative. The polyphony of opinions and ideas is designed to be mixed freely. There are, however, some broad overarching themes that link the individual texts together, which are briefly outlined below.

Participation, whether through global sound-system networks, post-digital street performance, or community theatre, is at the fulcrum of many of the projects and research.

A critical approach to design that foregrounds inclusivity is another recurrent theme. With examples from audio-description, soundscape design, and Machine Listening, we critique the extractive, racist, ableist, normative, market-led focus of so much of the designed world.

Space and spatiality in a relational, Lefebvrian[8] fashion of interconnectedness are other themes visited in this collection. Where there is space, there is sound. Where there is socio-spatial justice,[9] there is sonic justice.

All the contributions in this book were written between 2021 and 2025. Many offer critical commentary on their chosen subjects and a reflection on urgent global and contingent issues at the time of writing. Some of these issues include Israel's genocide in Gaza, the global COVID-19 pandemic, the rise of fascism in Latin America, Europe, and the United States, and the implementation of Artificial Intelligence and Machine Learning into much of our personal and institutional lives.

Cecilia Wee

The Grenfell disaster in West London on 17 June 2017, which killed 72 people, was one of the worst modern disasters in the UK. The horrific loss of life due to negligence by developers and the local council exposed socio-political relations, class, and racialisation in the area and the country in general. One of the most powerful actions that came out in the wake of the fire was the regular Grenfell Silent Marches. Adding to the ecology of impedance from local communities, these marches showed the potential for people to come together in an act of silent defiance. The direct opposition through a deterritorialisation of the streets through the power of silence showed how sound, in a deliberate rejection of semantics, could articulate community and political injustice. Drawing on personal testimonies, writings on sonic geographies, community organising practices, and case studies, this chapter examines the sonic, spatial, economic, and political dynamics of the Notting Hill Carnival, community activism in the wake of the 2017 Grenfell Tower fire, and the practice-based arts and learning project *Our Community Inheritance*, initiated by the author and made with local communities.

Harshada Balasubramanian

Harsha Balasubramanian's chapter focuses on audio description, the sonic practice of giving live or recorded descriptions of visual elements of an experience, film, or play, and it ends with a manifesto for its use within design. Echoing calls from the design justice movement, Harsha's contribution asks for the terms of design to be set by individuals and communities and not by designers and industry. The potentialities of Harsha's work go way beyond this important growing area of sound design in that the methods she advocates have agency in reimaging standardised approaches to design in general. By placing those often most marginalised by design at the centre, this work shows the creative potential for the constant involvement of disability communities within design through an in-depth examination of this particular sonic practice. Drawing on feminist theory and practice, Harsha demonstrates how Audio Description necessitates a constant reflexive relationship between designer and user, a relationship that can reconfigure the standardised, hierarchical relationship between designer and user, so common in design.

Pedro JS Vieira de Oliveira and Shintaro Miyazaki

Pedro and Shinto's chapter exposes the bind between machine-listening and colonialism through its use as a mechanism of corporeal measurement in Germany. Focusing on the use of machine listening as a tool for immigration control, the work examines how the standardisation inherent in machine listening is linked to racist methods of control and treats voices as somehow separate from our bodies. They trace machine listening's lineage to early colonial and scientific practices, such as anthropometry, eugenics, and phonographic recording, used to define and police human difference. Fundamental to their work is a rejection of the notion that machine listening is more neutral or objective than human listening. Instead, they explain how machine listening is co-constitutive and shaped by social, cultural, and political structures.

Nathanael Kliebhan

Nathanael examines the exclusionary nature of socially and physically constructed white spaces. His work asks us to reimagine Eurocentric conceptions of space through an anti-colonial spatial discourse that follows a musical logic where sound is fluid, permeative, and transgressive of physical boundaries. Central to the work is Nathanael's notion of "perforative" practice, a conceptual and enabling creative tool used to resist racist cultural enclosures and challenge the violence of exclusionary spatial practices. His work speaks both of absence and abundance. Of silence and agitation. Demonstrating how the black sonic reveals and reconnects spheres and practices towards alternative spaces.

Mariana Carvalho, Laura Mello, Nirah Pomar, and Elif Gülin Soğuksu

Created during the global pandemic, *Speaking-Through* from 2021 is a polyphonic, feminist, networked performance. The project mixes online and in-real-life coming-togethers, connecting across languages, timezones, and continents to create a choral tapestry of solidarity. Initiated when Jair Bolsonaro was in government in Brazil and a similarly oppressive political system existed in Turkey under Recep Tayyip Erdoğan, the work explores the body as a resonant container, an instrument of sonic disobedience. The chapter reflects on how the project moves through digital feminine space, with a focus on listening with, rather than listening to. Centred around the voice, or voices, the work is situated in different urban locations, and the discussions patch into the project's relations to female experience of body, identity, migration, history, and media. The performances described reimagine the use of the ubiquitous tool of sound art, the loudspeaker, through bodily augmentation. The resultant polyphony of languages, bodies, and voices amplifies stories of marginalisation and oppression.

Brian D'Aquino

Brian's work involves ten years of research within sound system culture and his involvement with the renowned sound system *Channel One* and the *Sonic Street Technologies Research Group*. The group connects sound systems in locations including London, Latin America, Japan, and Australia, forming a body of entangled sonic knowledge. Building on the work of Julian Henriques (Henriques, 2011), which places the body and its connection to other bodies in the making and enjoyment of sound systems at the centre of theorising sound systems as assemblages of knowledge, Brian's chapter highlights the invaluable contribution of Caribbean sound systems to a wide range of musical genres and cultures. In particular, the chapter focuses on the DIY nature of sound system culture and the importance of tacit, situated knowledge, handed down through informal apprenticeships and other forms of making together. A community-centred approach, far removed from commercially driven solutionist approaches to sound design.

Rosie Ash

Rosie's chapter focuses on her work with a migrant women's group in London who come together to make music and listen to each other's stories. Through their experiences of leading music and listening activities with the group, Rosie uses the sonic to encounter a range of areas, including climate change, late capitalism, and the UK immigration system. Through their study of voice, sound, and space, Rosie critiques the historic systemic oppression and victimisation of female migrant communities within London and uses sound as a means of amplifying social tensions, with resonances for wider spatial and political struggles. The text uncovers how the transformation of individuals arriving in the UK into the category of asylum seekers enables political agendas to distort or fictionalise the lived, often traumatic, realities of these people.

Manal Massalha

In this chapter, Manal foregrounds listening, not merely as a sensory act, but as a critical, political, and embodied practice rooted in place. In the discussions that form the text, Manal advocates for a decolonial listening praxis that values diverse sonic presences and challenges dominant hierarchies. Through discussions on how soundscapes capture lived experience, environmental change, conviviality, and systemic violence, Manal links Hackney in London and the city of Lydd (Lod). In Hackney, the multiplicity of sounds, from birds and bikes to prayers and music, becomes a hopeful symbol of pluralistic coexistence. In Lydd, Palestinian hip-hop, particularly through Tamer Nafar, emerges as a means of amplifying marginalised voices and memories. Listening, for Manal, is both a political act and a tool for reimagining inclusive, equitable futures across geographies of trauma and community. Listening resists control and becomes a form of witnessing, of amplifying voices, memories, and presences that violent colonialism tries to suppress.

Matt Lewis

This chapter critiques the functional use of sound in urban and architectural design to serve commercial agendas, a practice the author terms "soundwashing". Drawing on sound studies and design justice principles, the piece argues that acoustic interventions, such as fountains and piped birdsong, are often used as masking tools which reinforce normative assumptions and exclusionary practices. These solutions often reduce sound to a visual or functional object, failing to engage with the complex, embodied, and contextual nature of listening. The work also critiques wider dominant, solutionist design approaches that ignore the diverse auditory needs of different individuals. Instead, Matt advocates for approaches rooted in community practices such as folk music and community theatre, where knowledge is shared and hierarchies are flattened.

Notes

1 *In Unsound: Undead*, Steve Goodman, Eleni Ikonika, and Toby Heyes explore the idea of vibrational continuum as a concept to rethink sound beyond the audible, a spectrum of vibrational phenomena that spans across and beyond human hearing. See Goodman, S., Heys, T. and Ikoniadou, E. (eds.) (2019) *AUDINT—Unsound: Undead*. Falmouth: Urbanomic.

2 The term zanana (زنانة) is an onomatopoeic term used by Palestinians in the Gaza Strip, referring to the buzzing noise from unmanned aerial vehicles (UAVS), specifically Israeli military drones.

3 *Radio Free Derry* and *Radio Free Belfast* were clandestine radio stations active during the early years of the Troubles in Northern Ireland. These stations played an important role in local communication, propaganda, and resistance during this time.

4 *The Voice of Free Algeria* and *Combatant Algeria*, were important radio stations during the Algerian War of Independence (1954–1962). They played a pivotal role in the National Liberation Front's (FLN) communication strategy, both within Algeria and internationally. Frantz Fanon, though not referring to these stations by name, writes extensively about the role of radio in anti-colonial struggle in his writings, especially in his text *A Dying Colonialism* (original French title: *L'An V de la Révolution Algérienne*, 1959).

5 *Radio Alice* was a pirate radio station that was part of the creative left Autonomist movement in 1970s Italy.

6 Pierre Schaeffer (1910–1995) was a French composer, engineer, and theorist. His work *Traité des objets musicaux* (1966), Schaeffer promotes the concept of the sound object (*objet sonore*), a theoretical and perceptual unit of sound that exists independently of its source. He also expounds a framework for listening that includes four modes to describe how we engage with sound in different contexts. See Schaeffer, P. (1966). *Traité des objets musicaux: Essai interdisciplines*. Paris: Éditions du Seuil. For English translation, see Kane, B. (2014). *Sound Unseen: Acousmatic Sound in Theory and Practice*. Oxford University Press.

7 In *The Tuning of the World* (1977), Murray Shaeffer sets the foundations for acoustic ecology and introduces the concept of the "soundscape". See Schafer, R.M. (1994) *The Soundscape: Our Sonic Environment and the Tuning of the World*. Rochester, VT: Destiny Books.

8 For Henri Lefebvre, spatial relations go far beyond the physical as they encompass socio-political relations and are never neutral. See Lefebvre, H. (1991) *The Production of Space*. Translated by D. Nicholson-Smith. Oxford: Blackwell.

9 Edward A Soja has been key in developing the concept of spatial justice, which links social justice to geographic space. For more, see Soja, E.W. (2010) *Seeking Spatial Justice*. Minneapolis: University of Minnesota Press.

References

Canton, W. (ed.), 2024. *Sonic Rebellions: Sound and Social Justice*. London: Focal Press, Taylor & Francis.

Chattopadhyay, B., 2022. *Sound Practices in the Global South*. London: Bloomsbury.

Chaves, R., and Lazzetta, F., 2019. *Making it Heard: A History of Brazilian Sound Art*. New York: Bloomsbury Academic.
Chaves, R. and Lazzetta, F., 2022. *Alto e bom som: uma história da arte sonora brasileira*. São Paulo: Edusp
Drever, J., 2022. *Aural Diversity*. London: Routledge.
Eidsheim, N.S., 2015. *Sensing Sound: Singing and Listening as Vibrational Practice*. Durham: Duke University Press.
Eidsheim, N.S., 2018. *The Race of Sound: Listening, Timbre, and Vocality in African American Music*. Durham: Duke University Press.
Goodman, S. 2010. *Sonic Warfare: Sound, Affect, and the Ecology of Fear*. Cambridge, MA: MIT Press.
Henriques, J., 2010. *Sonic Bodies: Reggae Sound Systems, Performance Techniques, and Ways of Knowing*. New York: Continuum.
Henriques, J. 2011. *Sonic Bodies: Reggae Sound Systems, Performance Techniques, and Ways of Knowing*. London and New York: Continuum.
LaBelle, B., 2018. *Sonic Agency: Sound and Emergent Forms of Resistance*. London: Goldsmiths Press.
LaBelle, B., 2020. *Acoustic Justice: Listening, Performativity, and the Work of Reorientation*. London: Bloomsbury Academic.
Mills, M., 2008. Deaf jam: Film, performance, and digital technologies. *Disability Studies Quarterly*, 28(4). https://dsq-sds.org/article/view/146
Schaeffer, P. 1977. *Traité des objets musicaux: essai interdisciplines*. Paris: Éditions du Seuil.
Sterne, J., 2024. *Diminished Faculties: A Political Phenomenology of Impairment*. Durham: Duke University Press.
Voegelin, S., 2017. *Sonic Possible Worlds: Hearing the Continuum of Sound*. 2nd ed. London: Bloomsbury.
Voegelin, S., 2018. *The political Possibility of Sound: Fragments of Listening*. London: Bloomsbury Academic.

1

SOUNDING AND LISTENING IN PLACE

Four Events in the Nottingdale Community After the Grenfell Tower Fire

Cecilia Wee

Introduction

The following text takes four sound events in the Nottingdale community[1] as its starting point: the rich sonic exuberance of the Notting Hill Carnival; screams for justice from the community, especially the bereaved, the survivors and the residents at Royal Borough of Kensington and Chelsea Council meetings relating to the fire at Grenfell Tower; the Grenfell silent walk as a silent moving wall of ritualistic solidarity; and the hums of agreement at a community workshop conversation during the practice-based arts and learning project *Our Community Inheritance*. By exploring these four sound events, I examine the politics of sounding in North Kensington, proposing that these events constitute different forms of sonic protest.

At the heart of this text is a case study on *Our Community Inheritance*, a creative action research project exploring more equitable resource circulation that I have initiated, made with local communities. Each edition of *Our Community Inheritance* brings together decolonial approaches to cultural production, knowledge from community organising, radical, community-centric approaches to fundraising, and liberatory understandings of economics as 'making home'. *Our Community Inheritance* creates hyperlocal archives that document understandings of diverse, intergenerational knowledge and wealth, and makes community-generated proposals that respond to what our communities need to thrive. The project works with communities on journeys to develop and enact decolonising community resources through dialogue, image-making, and community organising. Edition #1 of *Our Community Inheritance* was developed in collaboration with NHS Grenfell Health and Wellbeing Service, and Grenfell-impacted communities from July to December 2022.

DOI: 10.4324/9781003194453-2

Reflecting on my experiences in the Nottingdale community before, during, and after the project *Our Community Inheritance*, I consider that the Nottingdale community has been practicing forms of sonic resistance and justice through celebration, silence, and voice. Further, I propose that conversational spaces of activists and community organisers are arenas for sonic archiving and activism – where listening and conversation are vital means through which we protest and heal. I consider not only how sound has operated within the historical cultural contexts of the Nottingdale community but also examine experiences of sound after the Grenfell fire and how the local community continues to resist, protest, and build on lineages of political agency in spite of and in the wake of the disaster. I also propose that the project *Our Community Inheritance* offers a way to explore how emerging configurations of political autonomy can look and sound within a hyperlocal context.

My relationship with the Nottingdale and North Kensington community as an outsider and ally has developed over a number of years. Since 2018, a group of students and tutors at the Royal College of Art (RCA) White City, including myself, have organised community banner-making workshops for the anniversary of the fire at Grenfell Tower, which is just twenty mins walk from the RCA's White City campus. This activity was fundamental to how I grew a relationship with members of the Nottingdale community. In my research, experiences, and the time I spent, I have aimed above all to listen to and value what everyone has told me. Through this partial account, I hope to contribute to an archive and understanding of how sound moves in Nottingdale, an incredible, tight-knit, and supportive community that continues to call for justice over the killing of 72 residents in the Grenfell Tower fire on 14 June 2017, the UK's worst public housing disaster in modern history.

Methodology

This text draws on a small number of key sources, experiences in the Nottingdale community, and secondary research. I am honoured to have been able to spend time and develop a friendship with Moyra Samuels, local resident, community activist, former teacher, and currently Community Connector for NHS Grenfell Health and Wellbeing Service. Moyra is an active and extremely well-connected member of the community, a key link between the RCA and Nottingdale, and was also absolutely vital to my project, *Our Community Inheritance*. Moyra was a primary informant for my work, advocating for my collaboration with NHS Grenfell Health and Wellbeing Service and inviting participants to *Our Community Inheritance*.

An interview with Moyra Samuels and Wilf Walker was also significant to my primary research. Wilf Walker was a participant in *Our Community*

Inheritance. He has been a local resident of Notting Hill for approximately 60 years, was an organiser of Carnival, and was central to the revival of musical culture in the 1970s and 1980s as a key promoter at Acklam Hall. Excerpts from my interviews with Moyra and Wilf, my reflections on working with the Nottingdale community, and my experience of the Grenfell silent walk are also part of this text.

Testimonies and oral histories from the North Kensington community, such as found in the volume *The Frontline: A Story of Struggle, Resistance and Black Identity in Notting Hill* (Blagrove Jr, 2022), edited by local activist Ishmahil Blagrove Jr, have influenced and shaped this work. Additionally, as well as drawing on literature that examines the spatial and political aspects of sound (Henriques, 2011) and the social and community dynamics of dialogue and voice (AM Kanngieser, 2011), my work has been informed by ethnographic studies and analyses by researchers, including Flora Cornish (2021), on groups, activism, and campaigns initiated by the Grenfell-impacted communities.

Approach

My approach to this research stems from my *values* as a practitioner and researcher, as part of, and in solidarity with, historically underrepresented and marginalised communities who are simultaneously at the centre of movements towards equity and yet are also silenced. My modes of working and action are therefore underpinned and informed by the work of the international Design Justice Network, liberatory pedagogies as articulated by activists such as Paulo Freire, and a seminal report, 'Why am I Always Being Researched?' (Chicago Beyond, 2019) on power relationships between communities, funders, and researchers, written by the impact investment philanthropists Chicago Beyond.

Understanding myself as a participant embedded within ongoing systems of extraction, yet with some agency to contribute to social change, motivates my aim to develop horizontal relationships of mutuality, reciprocity, and commoning with those I work with. Hence, I use participatory action research as a multifaceted research methodology (Rappaport, 2020) that acknowledges and embraces experimentation, creative methods, solidarity with lived experience, democratic models, research-based action, and reflection with (rather than 'for') communities.

Positionality

Lifting up and valuing the experiences, knowledge, and skills of historically marginalised and underserved communities – specifically, individuals who identify as part of Global Ethnic Majorities, d/Deaf, disabled, and neurodivergent, LGBTQIA+, migrant, and working-class communities – is germane to this project and my research. Instead of framing these communities as being in deficit, or primarily as beneficiaries, my work aims to create spaces that facilitate dialogue and liberation, and enquires into embodied experiences of complexity, contradiction, and injustices of the worlds we live in. I develop spaces of sounding (Henriques, 2011, p. xviii) as relational and material activities that draw in and resonate with others, working to layer and weave unheard and new stories with both existing and new community connections, including non-arts professionals, community activists, and organisers as collaborators and equal voices.

In this sense, my work considers sounding and cultural production – particularly when working with and celebrating the viewpoints of historically marginalised and underserved communities – as a form of activism and futuring: a form of transformative, prefigurative politics (Boggs, 1977, p. 100), and science fiction that builds bridges towards worlds, environments, and societies that we want to live in, as we go.

Sounds Celebrated and Extracted from the Ground of The Grove

The Westway as Architecture of Control

Going west out of London, I remember the predictable stop-start traffic before the Westway rises up between shiny buildings, and the elevated flyover opens up towards Heathrow and the West of England. Cars above, people below – an image of dominating infrastructure that was so in contrast to the first time I went to 'The Grove' – the colloquial term for the area of North Kensington stretching from Portobello Road to Harrow Road (in the North) to Westway Sports Centre (in the West). I was part of a young people's music project in Southwark, and we were visiting a similar project on Acklam Road, which runs parallel to the Westway. That short encounter left a strong impression of intimate, inhabited streets, and a strong sense of community – all of which cannot be discerned from the roads above.

Constructed as part of a high-speed road network to manage traffic in and out of central London, the Westway was built between 1962 and 1970. It is an architectural backdrop of The Grove, dominating and moulding public space in North Kensington. Little consideration of environmental and social impact was given to its planning, whose route was determined as the easiest path between Paddington and the West Cross Route. This noisy, dirty, and disruptive construction process demolished homes, displaced thousands of families, and cut communities in North Kensington in two. Within the first few months of opening, the Westway brought noise and air pollution, and disruption, with 47,000 cars a day to the area.

More than 50 years since its completion, the long-term impacts of the Westway are ongoing poor quality environmental conditions, and poor health outcomes for communities living between and underneath the Westway. As is the case with most major roads in London (and across the UK), levels of Nitrogen Dioxide (NO_2) along the Westway regularly exceed EU annual limits. Despite protests from the local community about the dangerous levels of NO_2 and proper use of the land directly underneath and adjacent to the Westway, the Westway Sports Centre is sited directly underneath the flyover, as are pitches for outdoor sports, including football, cricket, and tennis. Local residents, including young children, are therefore forced to exercise under and near the Westway, undoubtedly impacting their health. Excessive road traffic noise from the Westway, producing a 'swishing white noise' (Select Committee on the Crossrail Bill, 2007, para 12100–12119), similarly has negative impacts on residents' health and wellbeing.

In the context of post-war urban planning in West London, the Westway was part of an architecture of environmental, political, and social control that had the impact of muting North Kensington communities with the 'white noise' of traffic – this had its corollary in the negative media coverage of the area's racialised, low-income residents. Arguably, the underlying causes of the Notting Hill race riots in 1958, 1976, and 1981 were drowned out by the white noise of the (white) media. For instance, police at the time of the 1958 riots downplayed the part of white 'working class mobs' in harming the local community, and this was only revealed 40 years later:

> Disturbances were overwhelmingly triggered by 300–400-strong 'Keep Britain White' mobs, many of them Teddy boys armed with iron bars, butcher's knives and weighted leather belts, who went 'n****-hunting' among the West Indian residents of Notting Hill and Notting Dale. The first night left five black men lying unconscious on the pavements of Notting Hill.
>
> *(Travis, 2002)*

Tuning the Streets Through Sound

As well as being the location of sports and leisure spaces, the Westway continues to function as a place for the community to converge, gather, and mount political resistance to dominant forms of power, and has a long history as a site for local campaigns and protests. The flyover is also a feature on the route of the Notting Hill Carnival, an international festival that has a strong community presence all year round. Reflecting on the importance of Carnival to the community, Moyra Samuels says, 'for us Carnival was better than Christmas. The anticipation and the sense of actual power. It would be our streets for a few days....It was the sound of resistance. It was a festival of the oppressed' (Samuels & Walker, 2023). The pages of the anthology *Frontline* are replete with testimonies of how protest against state-sanctioned oppression was integral to the development and ethos of Carnival. For example, Leslie Palmer, a key organiser of the Notting Hill Carnival since its inception in 1966, states, 'Ladbroke Grove has played a major part in the resistance to police brutality and inequality in Britain. It was the principal centre of Black resistance in the country' (Palmer, cited in Blagrove Jr, 2022, p. 229).

The power of Carnival that Moyra and Leslie Palmer refer to is both symbolic and literal, with speaker walls filling streets with the sounds of defiance and celebration. The network of sound systems at Carnival is the fundamental technology that projects community-built sonic power back onto the streets of The Grove. Julian Henriques, who has written extensively on the theory, practice, and politics of sound systems, describes them as,

> Sonic war machines, vehicles for cultural expression, vessels for identity and pleasure, economic engines, commercial ventures, instruments for musical production, institutions for artists' training, multimedia communication systems, test-beds for technological innovation, laboratories for sonic science, programme content providers for local commercial cable TV stations and ambassadors for Jamaican music abroad.
>
> *(Henriques, 2011, p. 3)*

Following Henriques, sound systems at Carnival can be understood as part of a tactical media ecology, where forms of media distribution are composed and built to intervene in resistance and in response to wider cultural norms, where subaltern and Other voices or positions are often variously marginalised or excluded. Sound systems in the context of Carnival function as an 'apparatus' (Henriques, 2011, p. 3) and a strategy of sonic activism that develops a community asset as part of grassroots organising strategies to resist oppression (Costanza-Chock, 2020, p. 92).

The visceral power to control crowds through sound systems was discussed in an interview with Wilf Walker, who described the impact of installing 20,000 watts of sound under the Westway flyover during Carnival. This event demonstrated the respect and esteem commanded by sound systems in their dual capacity to generate sonic 'livity' (Henriques, 2011, p. 7) and non-violently disperse crowds in a way that was not available to the police. Carnival's sound systems can be framed as community-led interventions to reclaim public space through sound:

> [we were] able to control the area and entertain the people. And so when we told them that it was actually over, the majority were prepared to leave because they'd been entertained for 48 hours. Whereas before we put the sound there, there was just confusion there.
>
> *(Samuels & Walker, 2023)*

Sounds of Economic Control

The power of the sonic realm is also recognised by authorities as a valuable force that can be extracted and monetised. Both the economic and cultural contributions of Carnival are undeniable, providing and supporting London's standing as a world-class cultural city and building positive narratives of Britain as a place with vibrant cultures. According to the Greater London Authority's most recent economic impact study of the Notting Hill Carnival, in 2002, 'Carnival generated approximately £93 million and supported the equivalent of 3,000 full-time jobs'. (Greater London Authority, 2004, p. 11) However, Wilf Walker notes that such economic benefit was hardly experienced by local residents, and that much of the labour that underpins Carnival is undertaken unpaid:

> There's a lot of money and none of it is really coming to us....We were so captured by the notion that we forgot to make the demands. We never spent any time saying, what's the Man putting in? How much is the State putting in? And the State was taking out so much every year.
>
> *(Samuels & Walker, 2023)*

Instead of ensuring long-term support for the racialised, low-income communities that are at the heart of Carnival, state institutions seemed to erect barriers in a number of ways. For instance, Wilf Walker states that because alcohol licences were not awarded to residents for the purposes of Carnival, this meant that for many years most of the roadside stalls at Carnival serving alcohol were having to operate illegally (Samuels & Walker, 2023). Furthermore, despite efforts by the community to push back against aggressive policing tactics, North Kensington has been the testing ground for insidious new forms of control and surveillance. For example, Carnival 2016 and 2017 saw the first public uses of automated facial recognition in the UK, despite the UK Biometrics Commissioner (Wiles, 2017) admitting that 'tests of facial matching for spotting individuals in large crowds have so far had very poor success'.

In these ways, we can observe layers of acoustic contingency and push-pull power dynamics between North Kensington communities and state authorities. It is only through the cooperation and assent of the local community and community organisers that Carnival can take place at all. At the same time, North Kensington communities have been denied the right to benefit economically from Carnival, whilst this 'right' is capitalised on by state authorities. Further, as we will hear later on, authorities in North Kensington have chosen to selectively listen to community voices they deem acceptable and ostensibly silence them when considering other community issues that do not relate to Carnival. The tolerance and legitimisation of sound as vital to Carnival is racialised, classed, politicised, and contingent on possibilities for cultural and economic extraction and social control by dominant powers.

Unwanted Sounds from Unruly People

In neighbourhoods – such as North Kensington – that are densely populated, the control and ownership of overlapping and competing 'acoustic arenas' (Blesser & Salter, 2007, p. 26) are often drawn along racial, class, and financial dividing lines. Overriding mainstream rhetoric about sound in public space centres around the ideal of quiet spaces, itself a privileged (Northern) European construct. As Nicola Di Croce describes, disturbance is shaped by those with economic and political power (Di Croce, 2022), an idea that is succinctly articulated by Wilf Walker (Samuels & Walker, 2023): 'they want you to turn the music down!'

Drawing on the work of David Harvey and Henri Lefebvre, Ascensión García Ruiz & Nigel South articulate the hegemony of quietness in the public domain as an example of the increasing domination of rights *in* the city over rights *to* the city:

> A right *to* the city can be defined as the citizen's right to access everything that the city offers and to do so as an expression of human coexistence.... The contrasting concept would be the matter of rights *in* the city – which would include various forms of protection regarding public use of city areas and districts.
>
> *(García Ruiz & South, 2018, p. 8)*

The increasing privatisation of public space, encouraged through marketing of new developments and the visual, aural, and physical 'cleaning up' of places for the purposes of gentrification, can be understood as part of contemporary strategies for aggressive control of public space relying on the privileged assumption that quiet places nurture healthy ears.

Indeed, in his essay 'Attuning to disturbance – Towards a Multi-species Sonic Ecology' Nicola Di Croce is more critical, noting that 'sounds' agency or affective capability is often exploited to reach economic and political targets – that are leading to a serious growth of social exclusion' (Di Croce, 2022). In 2021, Westminster City Council took out an injunction banning social gatherings in Maida Hill Market Square to the North of the Westway, citing noise complaints. The injunction was primarily directed at a group of older Caribbean men who gathered in the square for more than 10 years to socialise, play dominoes, cards, and backgammon. In 2022, Ernest Theophile took Westminster City Council to court over the injunction, arguing that it discriminated against Caribbean culture and won the case. Theophile said: 'The loneliness was one of the biggest factors as to why we gather there....We just want a place where we can socialise and play a few games [of dominoes] to pass the time' (Taylor, 2022).

The injunction against the dominoes group can be understood as an example of what Jennifer Lynn Stoever calls the disciplinary 'listening ear'

(Stoever, 2016, p. 14), which intervenes in the public construction of power relations through sound. Such an intervention is made possible through the social reinforcement of a 'sonic colour line' (Stoever, 2016), which deems racialised bodies as unruly and noisy. In this context, residents, particularly those who are working class, from Global Majority backgrounds and/or have English as an additional language, are pitted against the quiet, genteel image of the well-behaved middle class and wealthy property owners, which we see taking place as part of the privatisation of space in North Kensington. As Edward Soja explains, contemporary methods of spatial control 'create and sustain unjust urban geographies' (Soja, 2010, p. 48) for increased financial gain, where 'waves of privatisation have been flowing into formerly public arenas of all kinds, compromising freedoms of speech, association, and political expression' (Soja, 2010, p. 48). Perhaps one of the notable aspects of the case of the dominoes group is that they dared to challenge the injunction, where the council may have expected them to defer to the law and 'keep the peace'. In so doing, the case drew attention to the presence of a 'sonic colour line' and upheld the dominoes group's right *to* the city.

In other words, whilst the sounds of The Grove, and specifically Carnival have been tolerated and legitimised, we have also seen that it is the respectable voices in the community that are more easily heard, since 'more often fetishized as exotic or demonized as unassimilable, noise and loudness frequently function as aural substitutes for and markers of race' (Stoever, 2016, p. 13).

Sounding Silence and the Grenfell Silent Walk

On the Fifth Anniversary Silent Walk

14 June 2022. As we approach Silchester Road, the pace is already slowing. The crowd gathers together at the corner, between Kensington Aldridge Academy and the Notting Hill Methodist Church. Green, the symbolic colour of Grenfell memorialisation, is all around – on rosettes, banners, on our bodies. As groups form, there are subdued nods, hugs, and chatter. I look down the road, and there are thousands of people behind us, all the way to the end of the road. Stewards quietly usher us along so that we are ready to begin. Hush falls, guided by the stewards, quietly announcing the start of the walk.

We move slowly down Lancaster Road, stopping every few steps. All is still except for the wind, the sound of birds, and the whisper of steps. I become aware of every movement, every bump in the road, the air as it brushes past my ear, and past the placards in front of me, as the procession navigates each bend and intersection. People stop outside their front doors and outside their shops.

The ushers stop us every 50 metres as we shuffle towards the corner of Ladbroke Grove, and as we approach, the acoustic horizon expands with the open road. The silence is broken by clapping that can be heard in the distance, resounding under the Westway between Sainsbury's and Pret a Manger opposite Ladbroke Grove tube station.

This is the only point on the walk where silence and walking are interrupted. A group of firefighters, wearing Grenfell hero badges, stand to the right of the road, forming a guard of honour with their helmets in front of them. We clap, and the march comes to a halt. There are handshakes, bows, and long hugs of grief, sadness, pain, respect, and thanks.

It is the most overwhelmingly intense part of the Silent Walk, and the emotional impact of these gestures is a shock on each silent walk, even though you know it is coming. It feels like we are directing the multilayered feelings of injustice, anger, loss, helplessness, fear, gratitude, and love for the community into the silent hugs and sonic event of clapping. We clap with all our hearts, and then we are silent again.

The Grenfell silent walk is a regular community-organised action, commemorating lives lost in the fire at Grenfell Tower on 14 June 2017. The first silent walk on 19 June 2017 was called by the community group Justice 4 Grenfell in the absence of an official mourning period, national minute's silence, or any recognition of the fire as the worst housing disaster in British modern history. Large gatherings on the yearly and half-yearly anniversaries of the fire have brought up to 11,000 participants.

The first Grenfell silent walk provided a frame for holding the community's grief immediately after the fire. As well, this silence was a strategic way to convene people of all ages, all faiths, and none, from different parts of the community in the aftermath, in a way that was dignified, meditative, and peaceful. The complete collapse of support from the Council, Kensington and Chelsea Tenant Management Organisation (who were responsible for managing the building), police, public health, and other authorities in the days and weeks after the fire was layered upon the circumstances of systemic neglect that led to the fire and exacerbated tensions in the community. Local residents, bereaved, and survivors, who had been abandoned after the fire, were demanding answers and expressing their rage, despair, and trauma. In the midst of the utter failure of the Royal Borough of Kensington Council to account for missing people, communicate adequately with abandoned residents, or provide safe alternative housing, the screams and outpourings of anger from the community were a response to gross and systemic negligence.

Racialised Tropes

The Grenfell inquiry heard that before the Royal Borough of Kensington Council relinquished control of the area six days after the fire, the Council 'formally attacked [the bereaved families, survivors and residents] as somehow would-be leaders of revolt' (Opus 2, 2022, p. 8). Mindful of media representations of young people within the community as uncontrollable or violent, local activist Ishmahil Blagrove Jr who was part of the silent walk mobilisation, called for this act of coming together in silence, in an effort to protect young people and reduce the chances of the community being subjected to increased arrests and stops by the police. Police presence during this period was highly visible and often hostile, particularly towards racialised young people within the community. This experience of excessive policing was proven to be based on Islamophobic decisions taken by the authorities – the inquiry found evidence that the council and police wanted to send armed police into the area the day after the fire because of fear of rioting, which they thought would be led by Muslim residents (Apps, 2022a).

The decision to create a silent gathering was also a response to how racialised communities are habitually held to stricter standards of civility and the associations of respectability with whiteness. As Jennifer Lynn Stoever says of the police brutality leading to the death of US woman Sandra Bland in police custody in 2015, 'white authority figures continue to expect black people to perform more visible, overt, and extreme forms of compliance – through speech, vocal tone, eye contact, and physical behaviour – than they ask of white subjects' (Stoever, 2016, p. 2). With the Grenfell fire, the media played a significant role in ascribing negative stereotypes to members of the community, mirroring a colonial biopolitical project where loud, ungovernable racialised subjects need to prove themselves as quiet, grateful, and subservient to be worthy of public support.

Silence and Silencing

The silence of the silent walk can at once be seen as a proactive reclamation of racialised, low-income communities raising their voices, yet it is also born out of the fraught reality of being silenced by authorities when battling for accountability and justice:

> Moyra – the silent walk tends to end up in some kind of parallel process over the years, reflecting the fact that we have been asked to be silent, because we are demanding way too much accountability and we are making too much noise and we are asking too many political questions in Council meetings…so eventually, in some way the Silent Walk started to reflect some of those areas where we can mourn our dead, but we need to fight like fuck for the living.
>
> *(Samuels & Walker, 2023)*

The act of silent protest and the experience of being silenced in the context of the Grenfell aftermath were two sides of the same coin. Jennifer Lynn Stoever says that as part of how the sonic colour line operates, 'white supremacy has attempted to suppress, tune out, and wilfully misunderstand some sounds and their makers and histories'(Stoever, 2016, p. 6). Testimonies during the Grenfell inquiry highlighted how bereaved families, survivors, and residents' needs and calls not only went unheard, but that there was an active attempt by the authorities to evade responsibility, cause division and uncertainty, denigrate the views of the community as a whole, and discredit outspoken local activists and residents.

Collective Silence as a Constituting of Power

The incorporation of a minute's silence on each day of the Notting Hill Carnival 2017 was proposed as an act of respect, remembrance, and mourning of the 72 people who had been killed in the fire at Grenfell Tower. Its successful coordination was a testimony to the community's solidarity and determination to rally together. Given the number of people attending, the numerous sound systems, and the decentralised ethos of Carnival, coordinating silence posed a challenging task. Carnival organisers instructed the sound system leads to broadcast the announcement through the radios. Those silences at Carnival were especially powerful and significant since the community was unsure if Carnival 2017 would take place at all, amidst media speculations and police warnings that outrage about the Grenfell fire within the community may spark mass violence and riots. Given that articulating a unique sonic identity and ethos are vital with sound systems culture, coordinating a silence not only requires a process of negotiation and agreement by all sound systems participating in Carnival but also relies on what Julian Henriques describes as each crew's '*phronēsis*, or practical wisdom' (Henriques, 2011, p. xxvii) of commanding and holding the sonic space.

The Grenfell silent walk builds on other historic silent marches undertaken by communities of colour in the name of justice, in which silence acts as a cipher for non-violence. For instance, the historic Negro Silent Protest Parade, also known as the Silent Parade, called by the NAACP in July 1917, involved 10,000 African Americans silently walking down Fifth Avenue, New York, to protest violence and lynchings against African Americans. Similarly, in September 1968, an estimated 40,000 students and workers staged a silent march through Mexico City against right-wing and government repression, calling for greater democracy.

These precedents are notable as they subvert our aural expectations of a moving protest, where making as much noise as possible and moving together as one loud body are conventionally seen as markers of success. The silent march impresses upon us from the combined perspective of scale and being quiet – an activity that asks us to consciously coordinate how our often-noisy bodies move with others. Moreover, we can place the Grenfell silent walk within a historical lineage of peaceful, nonviolent protest by communities of colour, countering narratives about peaceful protests as the privileged domain of law-abiding white bodies. In this way, the Grenfell silent walk not only reclaims silence as a form of respect for the dead, a refusal to allow a 'reworking of demands for justice into soundbites that can be neutralised' (Jones, 2021, p. 59), it is also an instance of communities of colour mobilising as a mass of people to actively protest against white expectations of who has the right to make sound. It is an embodied and collective act of resistance that challenges the notion of communities of colour as rowdy troublemakers.

Collective Refusal

The silence of the Grenfell silent walk is an active choice, a collective practice where no words are sufficient in relation to the institutional negligence and cruelty of the Grenfell fire. From the community perspective, it is a coming together and ritual practice of meditative remembering, but also, as it faces the public and the street, it resists easy absorption – the silence remains difficult, palpable, and tangible. The silence of the silent walk is a collective refusal for the voice to be co-opted, for one voice to be legitimised over another.

Mladen Dolar describes the voice as 'the medium of democratisation of justice....democracy is a matter of immediacy, that is, of the voice; the ideal democracy would be the one where everybody could hear everybody else's voice' (Dolar, 2006, p. 109). Read alongside this, the active, collective choice for the complexity of the community's voices to be present but not deployed in the silent walk maintains the ritual aspect of witnessing but also resists bending to existing forms of democracy and justice that have been set out by the law and whiteness. The silent walk must therefore be seen as an act of protest against how the voice is controlled by what Althusser called Ideological State Apparatuses (Dolar, 2006, p. 107) – it is a protest against these manipulations and capture of voice by hegemonic power in light of the Grenfell disaster. These sonic bodies refuse to give their voice for it to be 'properly circumscribed, tamed, pacified' (Dolar, 2006, p. 113), which is how the voice is often put to work by Ideological State Apparatuses.

The Grenfell silent walk can be considered an instance of how sound works to participate in constituting power, by and of the community, in opposition and in contrast to how the Power of the state[2] has tuned out, muted, and silenced the community's concerns. This affirmative and active practice of remembering, and the embodied experience of making silence together, is a solidarity action, where the community's control and mobilisation of sound is enfolded into its organising and operating power.

In her book *Violent Ignorance* (Jones, 2021), Hannah Jones identifies how state-funded complacency, abandonment, and discrimination manifest and result in the deaths of migrants and of marginalised, minoritised communities.

> The Grenfell Silent Walk is powerful because of its sustained and dignified silence in the midst of noisy London streets. But this is not a passive silence or one without meaning or charge. It is a silence that cannot be easily ignored. Rather, the silence demonstrated both the determination for change and the unspeakability of the injustice of what happened to Grenfell residents....The power of the walk is in their ability to unite in determined solidarity expressed through their shared silence and consistency. They are not going away.
>
> *(Jones, 2021, p. 57)*

Like Peter Apps (2022b) and Flora Cornish (2021), Jones emphasises that although the community's silent refusal to speak in the moment of mass protest is political, in other words, of the *polis*, it is not the only response. As a regular act/practice of trans-sensory (Chion, 1994, p. 136) experience that brings people together to mark the event of the fire, the silent walk is part of a network of community actions and initiatives in service of remembering, holding space, calling for accountability, and building for a community's future.

Voices of the Community – Our Community Inheritance, a Case Study

Mutual aid was a term that rose to prominence during the COVID-19 crisis, when hyperlocal networks across the globe quickly increased to support communities at a time when many public services halted their in-person support. Flora Cornish documents in her research (Cornish, 2021) that the spirit and practice of mutual aid had long been a feature within the Nottingdale community, developing in response to a unique set of circumstances – namely, historic (and ongoing) social justice battles. Specifically, these were battles against racism, income, and class divisions; the failure of local authorities and statutory organisations to invest in and ensure liveable environments, and the challenge of sustaining the year-round web of initiatives needed to support the annual Notting Hill Carnival. Such circumstances continue to have a disproportionate impact on minoritised and low-income residents, resulting in a breakdown of trust in authorities who have failed to be accountable to those they serve. As Flora Cornish says of community projects in light of Grenfell, 'in insisting on caring for others' lives, activism succeeds in instantiating a caring world. Beyond hope and despair is the staying power of communities who value human life and solidarity and keep fighting for them' (Cornish, 2021, p. 293).

Within the Nottingdale and North Kensington community, mutual aid and community-organised initiatives were often instigated by local activists who identified a strong need. For instance, the Pepper Pot Centre – located under the Westway, one minute away from the silent walk route – is a community centre for the older West Indian and Caribbean community working to combat loneliness and isolation of the Windrush generation. The centre was established in 1981 by community activist Pansy Jeffrey. Inevitably, some of the community mobilisations have been short-lived and/or time-limited, whilst others have continued to have value for generations of residents.

Moyra Samuels corroborates that community organising multiplied exponentially in the wake of the fire at Grenfell and was indicative of the change in dynamics where 'the community had started to actually see each other again, and listen to each other' (Samuels & Walker, 2023). Spending time in the Nottingdale area, I'm struck not only by the proliferation of community initiatives offering healthy eating, mental health support, education, and wellbeing support for local residents, but also that these groups are also contexts for conversation, connection, and deepening relationships within the community. The role of such community spaces to meet and talk is emphasised by Amer Kanngieser, 'the spaces of meetings are spaces of elocution and acoustic communication steeped in codings, narratives and histories, where space, politics and the oral/aural produce one another' (Kanngieser, 2011, p. 348). Such co-creation of voice and space works within a context of intergenerational care (particularly for young people) and community (as opposed to individualised) actualisation in the Nottingdale community, and is the backdrop for my project *Our Community Inheritance.*

A Background to 'Our Community Inheritance'

> *Our Community Inheritance* created a rare environment of mutual respect between generations and people from different walks of life to hold each other's experiences within the community, gently disagree, practice endurance and engage in laughter, creativity and fun in the process. Through a strengths-based arts-led approach, participants gained self-esteem, increased confidence, and their attitudes and outlooks improved.

NHS Grenfell Health and Wellbeing Service

Our Community Inheritance is a creative action research and learning project exploring more equitable circulation of wealth and resource generation, which I initiated and am making with local communities.

Edition #1 of *Our Community Inheritance* (OCI) was created in collaboration with art, community, and health professionals from the NHS Grenfell Health & Wellbeing Service and Grenfell-impacted communities, including people who use the Dedicated Service.[3] The project was a five-month commission as part of the *Sunlight Doesn't Need a Pipeline* project, by Dr Dani Admiss, hosted by Stanley Picker Gallery, Kingston University, and the research was supported by a grant from Clore Fellowship and Arts and Humanities Research Council (Wee, 2023b). OCI built an activist conversational space for voicing shared interests, concerns, and disagreements. The process made space for recognition and acknowledgment of different positions and experiences resulting from chronic under-resourcing within the community, making visible the layers of hyperlocal, complex experience that are the long-term consequences and impacts of colonialist capital.

OCI builds on my ongoing research and interventions into the long-term impacts of the 2008 global financial crisis, alternative economics, complementary currencies, and alternative funding infrastructures for arts and culture. It is influenced by the work of many artists, designers, activists, and theorists who are working against the long-embedded tropes of white, male, hetero, able-bodied, neurotypical, middle-/upper-class power as exclusively shaping what constitutes wealth and resources in our communities.

For instance, Gargi Bhattacharyya's work shines a light on how contemporary capitalism – as a legacy of the British empire – impacts communities in different ways (Wee, 2022); Edgar Villenueva's concepts of 'money as medicine' (Villenueva, 2018) and decolonising wealth as 'closing the (racial) wealth gap' have been fundamental to my framing of giving and resourcing; JK Gibson-Graham's iceberg model of economy (Gibson-Graham, 2020) builds on Hazel Henderson's love economy layer cake to articulate forms of labour beyond official markets. Gibson-Graham discusses the labour, knowledge, and expertise that we and our communities give, which are not valued because they escape being accounted for within official economic measures and ciphers of value. Historical and contemporary experiments in community and complementary currencies and mutual credit systems have also sought to redraw the parameters of what we consider to be valuable, namely, our relationships and partnerships as humans.

Radical philanthropy initiatives such as The Giving Project (in the US), Resourcing Racial Justice, and the Edge Fund (in the UK) challenge the dominance of scarcity economics, and also centre and lift up historically underrepresented and marginalised communities. Similarly, the work of the Design Justice Network and equitable listening practices as championed by Faizullah and Khan (2020) aim to remake the circulation of power and agency between project managers (or funders) and 'beneficiaries' towards co-production and co-creation.

Artistic projects such as Kathrin Böhm's work on the Centre for Plausible Economies, the Radical Renewable Art and Activism Fund (now Glasgow Community Energy project), Raju Rage's Under/Valued Energetic Economy project, and the Arte Util project have also been important references in developing this project.

My approach resonates with Olúfẹ́mi Táíwò's constructive view of reparations as 'build[ing] a just world, where the places that inherited the liabilities of yesterday ought to shell out the most for constructing a better world' (Táíwò, 2022), a project of future-making, transformation, and praxis. Such a conception of reparations involves assessing current resources, analysing how and why problems or a lack within a place exist, and asking questions about what can be done to change the current flows of resource distribution.

These artists, designers, activists, and theorists provide insightful critiques of capital, land, and labour in its various forms, offering alternative ways of being, models, and visions that contribute to the creation of more just and inclusive communities. Moreover, their methods highlight the importance of listening and conversation as indispensable intellectual, artistic, and political practices.

The proliferation of interest in decolonising culture demonstrates a real need to create space – within and beyond academic spaces – to highlight and deconstruct the insidious ways in which colonial legacies continue to shape power divisions and ways of living. By reflecting, listening, and discussing experiences within hyperlocal community contexts, OCI works to contribute to initiatives resisting and protesting oppressions, identifying and envisioning alternative futures, and participating in a prefigurative politics (Figure 1.1).

How 'Our Community Inheritance' Worked

At the heart of the OCI collaboration with NHS Grenfell Health and Wellbeing Service were intergenerational, intersectional dialogue, and image-making workshops with Grenfell-impacted communities at ACAVA studios

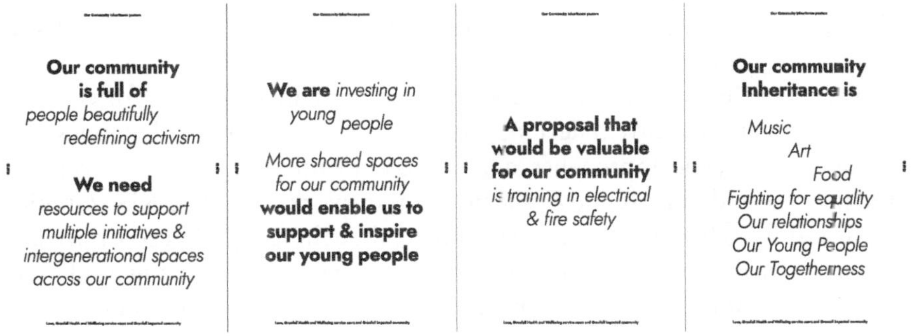

FIGURE 1.1 *A poster from the first Our Community Inheritance* workshop with Grenfell Health and Wellbeing Service, as part of Sunlight Doesn't Need a Pipeline.

and Morley College in North Kensington. Participants included market stall holders, artists, gardeners, community organisers, community health and social workers, young people from the Grenfell Health and Wellbeing Service User Involvement Team[4], and users of the Dedicated Service. Members of the community discussed and created alongside one another, exploring what intergenerational inheritance, in all its forms, means for the community. The aim was to think about all that we learn and gain from our cultures and heritage, across generations, to reframe our understandings of financial, spiritual, environmental, emotional, physical, and psychological health, wealth, and inheritance.

We invited workshop participants to share with us their stories and experiences of intergenerational knowledge, talent, and wealth in the community. We asked people about their superpowers and the knowledge, skills, and gifts they have learnt from older and younger generations, mapping these community assets. Participants made these visible and tangible through conversation and image-making, responding through mapping, writing, drawing, and model-making. This archival exercise served as a valuable form of reflection and celebration of the incredible skills and energy of the community.

In addition to these archival conversations, we invited participants to offer their proposals about what would be beautiful, inspiring, useful, and valuable for their community, building on the various community assets they had identified and exploring what they could bring to make these proposals happen.

There was a diversity of perspectives and experiences from participants, including older people, young people, and community activists. The discussion examined what climate justice means within the context of Nottingdale, North Kensington; local communities as self-organising knowledge hubs; the history of activism in the local area; struggles against inequality and racism spanning several generations; the long-term impacts of the Westway; the lack of good quality education; the lack of publicly accessible green spaces and play spaces; and the loss of affordable healthy food as a consequence of successive waves of gentrification.

The workshops provided an open and non-judgmental space for community activists to honestly reflect and think through community dynamics, different opinions, and approaches, as a means of identifying strengths, struggles, and needs as individuals and of the people they work with. The NHS Grenfell Health and Wellbeing Service team noted that the project's use of art and language encouraged participants to relate to their experiences in a different way. Participants reflected on how the local community has developed modes of resilience over generations, recognising the need to work together to support a broad spectrum of activities and initiatives for the community, affirming the importance of activism in every form. Alongside this, participants reflected that they wanted to ensure that future generations could inherit and benefit from conditions that are healthy, supportive, and safe. There was a strong sense that much of the community's activism and organising came from wanting to 'pay it forward', to support and invest in young people, by creating environments where young people can learn, experience, and thrive. Community organisers collectively acknowledged that even though their personal passions influence the focus of their projects, they held a shared objective of passing on 'skills for living' to other generations.

Drawing attention to and documenting what is inherited and valued within local communities – people, objects, places, vibes, events, and knowledge – these workshops worked with participatory budgeting frameworks to make small-scale projects that respond to what the community needs to thrive. The project's 'strengths-based approach' (Department of Health and Social Care, 2019) gave participants confidence and a positive vocabulary to speak about and accept contradictions and differences within the community in a non-confrontational way, allowing participants to reflect on and share their personal stories and life experiences and identify meaningful areas for community action and change through collective conversation, building capacity for community activists.

Simultaneously moving between different modalities and temporalities of dialogue and image-making opens up a multivocal and multivisual space for undervalued narratives, giving texture and ground to our experiences. OCI creates a light pedagogical frame and space of shared learning for communities to remember, to outline what we cannot write, and to envisage the new.

The workshops allowed participants to engage in critical thinking and active disagreements, whilst also successfully facilitating a valuable cross-cultural fertilisation of views, generating productive conversations between members of the community who might otherwise not connect with one another, who would usually not feel comfortable or confident to speak about their experiences in public. For instance, the workshops resulted in conversation between older Caribbean men and young Hijabi women, making space to articulate shared concerns and observations about members of their families and community, and jointly developing a community proposal. OCI enabled community participants to build experiments in participatory and inclusive community decision-making.

As a team, we were mindful about not replicating projects that already exist in the community, communicating clearly that projects with a clinical health need could be developed and supported through the NHS Grenfell Health and Wellbeing Service. As a time and resource-limited intervention, OCI does not aim to replace statutory services but rather supports the Grenfell-impacted communities to create more resources.

I presented the proposals for community projects with community participants and the NHS Grenfell Health and Wellbeing team at the *Sunlight Doesn't Need a Pipeline* festival at the Stanley Picker Gallery on 7 October

2022, alongside a series of objects made by workshop participants and the communities they work with. Each object told a story about the intergenerational knowledge and wealth that the Nottingdale community wanted to acknowledge. Through this presentation, the audience was also invited to reflect on and narrate their own story of intergenerational knowledge and wealth.

Edition #1 of OCI closed with another workshop, reconvening members of Grenfell-impacted communities to develop more proposals. After a five-month working process, Edition #1 of OCI identified several key projects for the community to take forward: an all-ages art club; an independent reflective space for community activists; safe, low-cost means for young people to access the natural environment; and a safe, reflective space for men to explore their emotions. The process also resulted in local activists and participants recognising ideas that they could bring forward in their own initiatives.

Drawing on participatory budgeting principles and frameworks,[5] community participants voted to support a proposal to create a safe, reflective space for men, which received £1000 seed funding from the fee that I was offered as part of my OCI commission for *Sunlight Doesn't Need a Pipeline*.

Edition #1 has now been handed back to the NHS Grenfell Health and Wellbeing Service, and Grenfell-impacted communities (Figure 1.2).

An Activist Conversational Space

The idea of community inheritance was extremely important for Grenfell-impacted communities, as communities working towards healing and justice, where formal justice has been lacking. *Our Community Inheritance* found alignment with NHS Grenfell Health and Wellbeing Service as we shared an aim to affirm the strengths, relationships, knowledge, and energy of the local community, as a means to heal, lift up, and create pathways for the future.

OCI created open and generative conversations with community members and local activists, broaching sensitive topics without reinforcing divisions. In particular, participants valued this space for community organisers to come together to discuss what is happening in their community and embraced the use of dialogue and image-making as opening up new paths of knowing and being. Such arts-based approaches to community conversations enable us to imagine, rehearse, and practice other ways of living, where articulating a range of perspectives on issues can often feel confrontational, combative, or subject to (self)censure.

Taking a plurivocal, multisensual approach to listening, OCI invites participants to engage in reflecting on the past, as well as zooming out to assess the present, project into the future, and share how they want to change where they live. As well as encouraging participants to time-travel, the approach allows participants to witness one another's concerns – live and in the arena

FIGURE 1.2 Illustration by Toya Walker, *Our Community Inheritance* workshop 28 November 2022.

of discussion, and once raised in this way, they are audible as *community, rather than private* concerns. OCI holds a space for the community as a means of de-privatising what Bruno Latour terms 'matters of concern' – 'whose import then will no longer be to debunk but to protect and to care, as Donna Haraway would put it' (Latour, 2004, p. 232) – by opening up collective consideration of situations that are intractable and complex.

OCI assumes that making 'oddkin' (Haraway, 2016, p. 4) is essential to life today, to (re)solve, work through, elaborate, and live with the huge issues that we face, that evidently cannot be sufficiently addressed through privatised, individual, and family unit approaches. We cannot live without oddkin – the mutual support aid groups of the COVID-19 pandemic were an example of this. OCI aims to create a frame for community members to collectively consider a locale in its contradictions, richness, problematics, and differences as individuals inhabiting shared space as well as groupings of people with common concerns and goals, uncovering that which is (in)visible, (in)active, present, absent, and otherwise unsaid.

OCI asks –

What needs in the community need to be uttered?
What are you hearing from those not present?
What legacies are unheard and still have resonance now?
What has been expressed that needs to be activated?
What are the echoes in the community today?

As an activist project, OCI aims to enable a multiplicity of voices to be heard and valued, activating our potential as individuals and a community. It is a contrapuntal unfolding of community dynamics and an opportunity for collective listening that is counter to the isolation and disconnection of how problems are often portrayed and understood.

Conclusion

Sound and silence have clearly shaped and marked various forms of collective action and solidarity within the Nottingdale community, particularly in the wake of the Grenfell fire. The community's public calls for justice after the fire are a result of the catastrophic failure of governments and authorities to provide safe, liveable environments for people in the Nottingdale community. This is symptomatic of wider failures of British governments, including climate destruction as a result of colonial pillage; migration due to unsustainable economic conditions and untenable habitats, and ghettoisation of swathes of the population, particularly those racialised as non-white and from low-income backgrounds due to gentrification and rising housing costs. Cultures of institutional neglect by authorities who were tasked with providing safe environments for residents, including vulnerable people in their charge, are layered upon complex histories and dynamics of classism, racism, and migration in these neighbourhoods of West London.

The active silence of the Grenfell silent walk should not have been necessary. It was the consequence of conscious, violently negligent silencing of minoritised communities, and the result of selective hearing by those who have the privilege of surrounding themselves with other types of noise, namely, the noise of capital, respectability, and power. The injustice of the Grenfell fire brings racial, economic, and climate justice home to us in the urban Global North, necessarily framing the community's use and interaction with sound as politicised.

For the Nottingdale and wider North Kensington community, the project of recording and archiving through and with sound, the community's multiple sonic, oral, and aural histories is significant. Questions of community archiving and memory-making are particularly important for racialised North Kensington communities,[6] whose stories have been historically neglected, and yet they have repeatedly been under media attention as instigators of social upheaval. As Barry Blesser and Linda-Ruth Salter note (2007, p. 17), 'aural architectural heritage' is difficult to archive and preserve on an everyday basis, due to how aural memory works and the temporal nature of sound. The aural architectural heritage of The Grove encompasses not only the Westway's deafening infrastructure from above *but also* the echoes of multiple race riots *and* the bass-heavy sound systems of the Notting Hill Carnival *and* the haunting quiet dignity of the Grenfell silent walk *and* protestors storming Kensington and Chelsea town hall demanding justice for Grenfell bereaved, survivors, and residents.

We are invited into remembering and connection through experiencing and noticing these rhythms, patterns, repetitions, and other functions of the *sonic* (as opposed to *representational*) logos – calling on Julian Henriques' (2011, p. 249) interpretation of remembering as 'an act of putting things back together, that is, of recognizing their connection….In this way connection is the reality and separation is the illusion that remembering dispels. Re-membering thus has its opposite in dis-membering, that is, tearing things apart'.

Sonic practices of remembering, like those embedded in the project *Our Community Inheritance*, are connective and relational practices, allowing us to make sense through sharing our views, our words, and perspectives. By focusing on listening, as an active and relational sound practice, *Our Community Inheritance* recognises and reiterates interdependency as a key value informing identity in the Nottingdale community and aims for a decolonial sharing of knowledge production that resists and goes beyond the written word. Indeed, decolonisation cannot be just about narratives of institutions; it is above all about our communities, everyday narratives, experiences, and relationships: 'the reclamation of resources, visibility and voice is a decolonial process of refusing the paternalism of organisational logics. What forms of resistance and refusal can we practice that simultaneously enable us to thrive?' (Wee, 2023b, p. 4).

Our Community Inheritance therefore proposes, as simultaneous critique and science fiction, that despite the fact our communities are at breaking point due to the multiple crises of capitalism, we – community members and their collaborators – also have the skills, power, experience, and resources within our reach to move towards healing and transformation. Moreover, such community listening and dialogue spaces affirm the power of the neighbourhood or community as a unit for organising, rather than reverting to the European heteronormative, able-bodied, middle-class convention of the family as a unit for organising.

The sound events and practices I have detailed are forms of sonic protest and resistance within the North Kensington community that speak of sound's capacity and contribution in helping build transformational and more just futures, which I consider the ultimate goals of protest.

Notes

1 The Nottingdale area is a community and electoral ward in North Kensington that includes the site of Grenfell Tower and is part of the Royal Borough of Kensington and Chelsea https://www.rbkc.gov.uk/pdf/Notting_Dale_Poll_District.pdf.
2 For an elaboration on the difference between power and Power as developed by Antonio Negri, see Michael Hardt, 'Translators Foreword: The Anatomy of Power', in Antonio Negri, ed., *The Savage Anomaly*, University of Minnesota Press, 1991, p. xiii.
3 The NHS Dedicated Service was designed with the bereaved and survivors of the Grenfell Tower fire and Grenfell Walk, to support and coordinate access to emotional and physical wellbeing health services https://grenfellwellbeing.cnwl.nhs.uk.
4 'The Service User Involvement Team was formed in 2018, and consists of Adult Service User Consultants, Young People Service Representatives, and clinicians who are engaged in a process of coproduction in the design and delivery of GHWS'. https://grenfellwellbeing.cnwl.nhs.uk/about/service-user-consultants.
5 *Our Community Inheritance* essentially operated a closed collective model of participatory grantmaking, see Hannah Paterson, 'Grassroots Grantmaking: embedding participatory approaches in funding', A Winston Churchill Fellowship report, 2020, p. 20 https://hannahpatersoncom.files.wordpress.com/2020/11/grassroots-grantmaking-embedding-participatory-approaches-in-funding.pdf.
6 North Kensington is, 'the most diverse area of Kensington and Chelsea, with a 36% BAME population', see NHS Grenfell Health and Wellbeing Service, 'About Us, North Kensington Community' https://www.grenfell.nhs.uk/who-are-we/north-kensington-area.

Bibliography

Apps, Peter, 'Islamophobic police document predicted "crime and disorder" after Grenfell fire due to "Muslim background" of victims', *Inside Housing*,

27 June 2022a https://www.insidehousing.co.uk/news/islamophobic-police-document-predicted-crime-and-disorder-after-grenfell-fire-due-to-muslim-background-of-victims-76262

Apps, Peter, *Show Me the Bodies: How We Let Grenfell Happen*, Oneworld Publications, 2022b.

BBC, 'Grenfell Tower fire: Silent walk marks first anniversary', 2018. https://www.bbc.co.uk/news/uk-44475913

Blagrove, Ishmahil Jnr (ed.), *The Frontline; A Story of Struggle, Resistance and Black Identity in Notting Hill*, Rice and Peas Press, 2022.

Blesser, Barry and Linda-Ruth Salter. *Spaces Speak Are You Listening?: Experiencing Aural Architecture*. MIT Press, 2007.

Boggs, Carl, 'Marxism, prefigurative communism, and the problem of workers' control', *Radical America* 11:6, 1977, pp. 99–122.

Bulley, Dan, Jenny Edkins and Nadine El-Enany (eds), *After Grenfell - Violence, Resistance and Response*, 2019.

Chicago Beyond, 'Why am I always being researched?' 2019. https://chicagobeyond.org/wp-content/uploads/2019/05/ChicagoBeyond_2019Guidebook.pdf

Chion, Michel, *Audio Vision*, Columbia University Press, 1994.

Cornish, Flora, 'Grenfell changes everything? Activism beyond hope and despair', *Critical Public Health* 31:3, 2021, pp. 293–305. https://www.tandfonline.com/doi/full/10.1080/09581596.2020.1869184

Costanza-Chock, Sasha, *Design Justice: Community-Led Practices to Build the Worlds We Need*, MIT Press, 2020.

Department of Health and Social Care, 'Strengths-based approach: Practice framework and practice handbook', 2019. https://assets.publishing.service.gov.uk/media/5c62ae87ed915d04446a5739/stengths-based-approach-practice-framework-and-handbook.pdf

Di Croce, Nicola. 'Attuning to disturbance. Towards a multi-species sonic ecology', *Sonic Urbanism: Listening to Non-Human Life, Theatrum Mundi*, 2022. https://theatrum-mundi.org/library/attuning-to-disturbance-towards-a-multi-species-sonic-ecology/

Dolar, Mladen, *A Voice and Nothing More*, MIT Press, 2006.

Drever, J.L., 2019. 'Primacy of the Ear'–But Whose Ear?: The case for aural diversity in sonic arts practice and discourse. *Organised Sound*, 24 (1), pp. 85–95.

Faizullah, Nusrat and Farzana Khan, 'Who has gained? Moving from extractive to equitable listening', *The Listening Fund*, 2020. https://www.thelisteningfund.org/wp-content/uploads/2021/02/who-has-gained-the-listening-fund-extract1.pdf

Ferrington, G., 1994. Audio design: Creating multi-sensory images for the mind. *Journal of Visual Literacy*, 14 (1), pp. 61–67.

Freire, Paulo, *Pedagogy of the Oppressed*, Penguin, 1972.

Fry, T., Stewart, S., Dilnot, C., 2015. *Design and the Question of History*. London: Bloomsbury.

Fryer, L. E., 2016. *An introduction to audio description: A practical guide*. London: Routledge.

Furniss, L., 2020. Beyond discipline: evolving design practice and design education in the twenty-first century. *Architecture_MPS*, 18 (1), pp. 1–17.

García Ruiz, Ascensión and Nigel South, 'Surrounded by sound: Noise, rights and environments', *Crime, Media, Culture*, 15:1, 2018, pp. 125–141. https://journals.sagepub.com/doi/10.1177/1741659017751223

Gibson-Graham, J.K. (ed.), *The Handbook of Diverse Economies*, Edward Elgar, 2020.

Greater London Authority, 'Notting Hill Carnival - A strategic review', 2004. https://www.london.gov.uk/sites/default/files/1782_-_strategic_review_2004.pdf

Grosz, E., 1999. *Becomings*. Ithaca: Cornell University Press.

Haraway, Donna, *Staying with the Trouble: Making Kin in the Chthulucene (Experimental Futures)*, Duke University Press Books, 2016.
Henriques, Julian, *Sonic Bodies - Reggae Sound Systems, Performance Techniques, and Ways of Knowing*, Continuum, 2011.
Jones, Hannah, *Violent Ignorance: Confronting Racism and Migration Control*, Bloomsbury Academic, 2021.
Kanngieser, A.M. 'A sonic geography of the voice: Towards an affective politics', *Progress in Human Geography* 36:3, 2011, pp.336–353.
LaBelle, B., 2020. Towards acoustic justice. *Law Text Culture*, *24*, pp. 550–572.
Latour, Bruno, 'Why has critique run out of steam? From matters of fact to matters of concern', *Critical Inquiry* 30:Winter, 2004, pp. 225–248.
Mazé, R., 2017. Design and the Future: Temporal Politics of 'Making a Difference'. In R. Smith, K. Vangkilde, M. Kjersgaard, T. Otto, J. Halse, and T. Binder (Eds.), *Design Anthropological Futures*, pp. 37–54, London: Bloomsbury.
Negri, Antonio, *The Savage Anomaly*, University of Minnesota Press, 1991.
Opus 2, 'Official court reporters, Grenfell Tower inquiry', *RT Day 263*, 11 April 2022 https://assets.grenfelltowerinquiry.org.uk/documents/transcript/Transcript%2011%20April%202022.pdf
Pain, Rachel, Geoff Whitman, David Milledge and Lune Rivers Trust, *Participatory Action Research Toolkit*, Durham University, 2012 https://localtrust.org.uk/wp-content/uploads/2019/03/local_trust_par_toolkit.pdf
Paterson, Hannah, 'Grassroots Grantmaking: embedding participatory approaches in funding', *A Winston Churchill Fellowship Report*, 2020. https://hannahpatersoncom.files.wordpress.com/2020/11/grassroots-grantmaking-embedding-participatory-approaches-in-funding.pdf
Rai, Amit S. and Will Essilfie, 'This is not 'interesting' research: Authentically co-creating participatory action research in UK's post-covid creative industries', *Conjunctions - Transdisciplinary Journal of Cultural Participation* 10:1, 2023. https://tidsskrift.dk/tcp/article/view/145681
Rappaport, Joanne, *Cowards Don't Make History: Orlando Fals Borda and the Origins of Participatory Action Research*, Duke University Press, 2020.
Remael, A. and Vercauteren, G., 2007. Audio describing the exposition phase of films. Teaching students what to choose. *Trans: Revista de traductologia*, 11, PP. 73–93.
Sacks, O., 1998. *The Man Who Mistook His Wife for a Hat and Other Clinical Tales*. New York: Touchstone.
Samuels, Moyra and Wilf Walker, Interview by Cecilia Wee [Zoom], 24 May 2023.
Select Committee on the Cross Rail Bill Minutes of Evidence, Examination of Witnesses (Questions 12100–12119), 2007 https://publications.parliament.uk/pa/ld200708/ldselect/ldcross/112/8050707.htm
Smith, R.C., Vangkilde, K.T., Otto, T., Kjaersgaard, M.G., Halse, J., and Binder, T. (Eds.), 2016. *Design anthropological futures*. London: Bloomsbury Publishing.
Soja, Edward, *Seeking Spatial Justice*, University of Minnesota Press, 2010.
Stoever, Jennifer Lynn, *The Sonic Color Line - Race and the Cultural Politics of Listening*, NYU Press, 2016.
Táíwò, Olúfẹ́mi, 'The constructive view of reparations and philosophical frameworks of global justice', *Georgetown Journal of International Affairs* 2022. https://gjia.georgetown.edu/2022/12/07/the-constructive-view-of-reparations-and-philosophical-frameworks-of-global-justice-with-professor-olufemi-o-taiwo/
Taylor, Harry, 'Dominoes player wins case over "racist" noise ban in London square', *The Guardian*, 14 May 2022. https://www.theguardian.com/uk-news/2022/may/14/dominoes-player-wins-case-over-racist-noise-ban-in-london-square

Tekin, Selin and John Drury, 'Silent Walk as a street mobilization: Campaigning following the Grenfell Tower fire', *Journal of Community & Applied Social Psychology* 31:4, 2021, pp. 425–437. https://onlinelibrary.wiley.com/doi/10.1002/casp.2521

Thackara, J., 1988. *Design after modernism.* New York: Thames and Hudson.

Travis, Alan, 'After 44 years secret papers reveal truth about five nights of violence in Notting Hill', *The Guardian*, 24 August 2002. https://www.theguardian.com/uk/2002/aug/24/artsandhumanities.nottinghillcarnival2002#:~:text=While%20senior%20officers%20tried%20to,boys%20armed%20with%20iron%20bars%2C

Villenueva, Edgar, *Decolonising Wealth - indigenous Wisdom to Heal Divides and Restore Balance*, Berrett-Koehler, 2018.

Wee, Cecilia. 'Livability - A conversation between Gargi Bhattacharyya and Cecilia Wee', *Sunlight Doesn't Need a Pipeline Newsletter*, 2022. https://sunlightdoesntneedapipeline.substack.com/p/livability

Wee, Cecilia. 'Casting spells', *Producing liveness in interesting times, the uncultured*, 2023a. https://www.the-uncultured.com/uploads/1/3/2/2/132249368/the_uncultured_-_pliit_2_casting_spells_v4.pdf

Wee, Cecilia, 'The role of cultural leadership in decolonising community wealth', *Clore Fellowship and Arts and Humanities Research Council*, 2023b. https://www.cloreleadership.org/research/the-role-of-cultural-leadership-in-decolonising-community-wealth/

Weidman, A., 2014. Anthropology and voice. *Annual Review of Anthropology*, 43 (1), pp. 37–51.

Wiles, Paul and Biometrics Commissioner, 'Metropolitan police's use of facial recognition technology at the Notting Hill Carnival', 2017. https://www.gov.uk/government/news/metropolitan-polices-use-of-facial-recognition-technology-at-the-notting-hill-carnival-2017

2

A MANIFESTO FOR INTEGRATING AUDIO DESCRIPTION IN DESIGN PRACTICE

Harshadha Balasubramanian

Introduction

This chapter bolsters calls to integrate access in design practice by alerting designers to the critical reflexivity that could be afforded by doing so. Conventionally, access is introduced as a retrofit solution, a functional adjustment intended to accommodate those who could not have otherwise experienced the designed service or product. Disability communities, especially rights movements, have historically been the locus of many political and technical battles to make access part of mandatory considerations in design (Williamson, 2012). However, the present chapter refuses to treat access as a stable, predefined objective or as select instances where the barriers confronting certain communities are removed (Hamraie and Fritsch, 2019). I follow Critical Access Studies scholars who acknowledge how disabled people engaging in access work find themselves continually critiquing and reimagining how society has been built (Jackson, 2018). Transplanting such access practices without regard for their complex histories or their communities would constitute appropriation. Instead, my proposition is to adapt access practices in a way that facilitates critical, disability-led intervention in design.

The specific contribution here is to argue that Audio Description (AD), an aural, participatory methodology shaped in the disability arts context, can inform designers' reflections and bend their practice closer to justice. AD traditionally involves verbalising visual details in media content for audiences who may be accessing it non-visually. I invite everyone partaking in the design process to describe their experiences, whether visual or not, including what is being designed, the tools used, their surroundings, and any other

DOI: 10.4324/9781003194453-3

phenomena that emerge as appropriate. These live, improvised, co-created descriptions differ from standard verbalisations of designers' encounters, as AD practice requires describers to capture how ideas have been materialised in a way that their listeners understand, an intersubjective negotiation that offers distinct possibilities for critical reflection amongst collaborators in the design process. Hence, the chapter hopes to reinvigorate interest in verbal methodologies at a time when non-verbal techniques are garnering appeal (Delle Monache et al., 2018).

In the ensuing pages, an initial section historicises AD, outlining the trajectory that has revealed its suitability to the present agenda. Three proposals are then made for aspects of AD that could be adapted and implemented in design practice. First, I contend that AD will enrich reflexive practices in design. Although reflective practice already occurs in design (Pihkala and Karasti, 2016), I suggest that these are often led by designers and conducted along prespecified categories of difference. Taking up feminist formulations of reflexivity that recognise power relations in design, namely Lucy Suchman's theorisation of "located accountability" (Suchman, 2002), I suggest that AD affords designers insight into how their knowledge may be situated, as well as how this situatedness makes them accountable in specific ways. My second argument claims that, while AD does not advocate for specific futures, it orchestrates a method for future-making that is reflexive of how this discovery of possibilities takes place. Third, I attend to how speech acts are increasingly endorsed as part of social justice goals in design practice without designers being sensitised to the role of the voice. I propose that designers should harness a describer's sensitivity for how material and symbolic dimensions of the voice mediate speaking and listening.

Having laid this ground, the chapter then provides a short manifesto which imagines what deploying AD in design practice could be, along with potential principles. I offer this cautiously, given the diversity that characterises what designers do and why. Nonetheless, readers may find that there are particularly significant implications for practices that struggle to be verbalised, such as sound design. Since sound designers have mostly developed audio for moving image, there has been limited rhetoric to discuss sound in its own terms (Collins, 2020, p. 11), a challenge which AD could arguably help to address.

Before proceeding, it is essential to position my upcoming treatment of AD theory and practice. I am approaching the topic as someone who uses the access provisions made possible by AD. I identify as blind, and while this does not solely steer my attitudes to AD, it has made me conscious of how such disability community practices are marginalised and of the implications for members of these communities. Although I would commend increased acceptance of AD as an accessibility measure, I am also invested in discovering ways in which interventions from disability communities can transform normative practices, rather than simply run parallel to them. It is this stance that has compelled me to adopt the usage of the terms blind and partially blind, phrases put forward in Hannah Thompson's advancement of "Blindness Gain", like "Deaf Gain", refuses to define disabled people by their relationship to able-bodied people (Chottin and Thompson, 2021). My critical awareness of AD draws upon having researched and collaborated with theatre audio describers in the West End, which gave me first-hand experience of the everyday discursive and embodied practices adopted by audio describers. By stating how my proposals are politically situated, I hope to have commenced the same journey of self-positioning that I am asking designers and their partners to embark on.

Audio Description as an Access Practice

AD practice has historically sought to selectively verbalise the "visual" features in media content so as to create access for blind and partially blind audiences, as well as those who may experience situational sight loss (Braun, 2008). Accordingly, AD practice is cross-modal, transferring meaning from content that is seen to content that is written and spoken out loud. Although most popular in film, television programmes, theatre, museums, and galleries, there are increasing exemplars that are making a case for using AD as an essential provision to enhance non-visual access in classroom learning, immersive technology, and games (Hughes et al., 2020). Audio describers develop and record scripted descriptions that must be heard in the gaps left by the auditory elements of the media content being described. In live events, these scripts are adapted to reflect any modifications that arise when the media content is delivered. While AD in much prerecorded content is transmitted through the same speakers as the rest of the audio output, the audiences experiencing AD at live events receive descriptions discretely via Bluetooth-connected headphones (Udo and Fels, 2010). The increasing prevalence of touch tours permits audiences to supplement any understanding developed through descriptions by touching objects, including museum artefacts (Hutchinson and Eardley, 2020) and theatre sets and props (Di Giovanni, 2018).

Despite recognising the multiple sensory aspects that constitute interaction, AD training has traditionally urged describers to attune their awareness to verbalising details that are exclusively visual and not materialised in any other way. Writing from the American context, where AD's formalisation took root, Joel Snyder prioritises observation as the first skill to master, contending: "The best audio describer is sometimes referred to as a 'verbal camera lens', able to objectively recount visual aspects of an exhibition or audiovisual programme" (Snyder, 2005, p. 195). This conceptualisation of AD as a transference of meaning has enabled its association with translation studies (Ibáñez and Vermeulen, 2013), where there have been similar debates about the extent to which translators should accurately reproduce the source content when trying to make it relevant for different target audiences (Romero-Fresco and Chaume, 2022). Just as translators take their primary guidance from the source being translated, audio describers concentrate on the media content they are making accessible, with rarely any input from its original producers. Describers need to develop an in-depth understanding of the medium of the source content and how it affects meaning, as well as how the medium through which the description will be received will influence meaning as it is conveyed (Mazur and Chmiel, 2021). This includes being able to recognise conventions specific to particular media which will be familiar to its intended audience, such as "crucial filmic clues meant for the original target public" (Remael and Vercauteren, 2007, p. 75), and being able to accommodate these in descriptions so that AD users can discover them too.

Unlike in translation, however, listeners are not always completely relying on AD when making inferences. In audiovisual content like theatre and film, dialogue, sound effects, and, where available, fellow audiences' reactions may give AD users vital clues as to what might be happening or alert them to the mood of the piece. For instance, in a classroom scene where a teacher has asked students to step forward, and listeners hear an audible step, it could be sufficient for describers to say a name, letting AD users deduce that this person made the step forward. When describers repeat an observation that AD users have already discerned, this has been said to risk encroaching on listeners' freedom to cast their own interpretations (Snyder, 2005). Instead of translating all visuals into words directly, describers have to dynamically respond to how listeners may variously leverage different ways of perceiving depending on the modalities that surface at any given time during their encounter with the media content. In addition to being concerned with the media of images and words, describers are compelled to note how their listeners' experiences, and hence the construction of meaning, are mediated by multiple modalities. The process of compiling AD, therefore, involves reflecting on the socio-material relations that inform how knowledge emerges in a given context.

Having considered the conditions under which listeners come to know, audio describers endeavour to anticipate how AD users may draw interpretations in situ, seeking to support listeners in this interpretive process rather than inundating them with information. In a podcast episode I produced following my research in London's West End, theatre audio describers explained how they did not aim to provide AD users with an accurate rendering of visual details or emulate the related conception of a "sighted person's experience". Indeed, they contested the very notion of anyone having a singular experience of a performance that can be reproduced. Roz Chalmers, who at the time described plays at the National Theatre, explained: "It's really a question of making sure that the person has an opportunity to engage with the show in whichever way they want to engage with the show" (Balasubramanian, 2017). Rather than upholding a "right" way to experience the show and inferring meaning on listeners' behalf, describers strive to ensure that AD users have the autonomy to attend to different elements of the production, allowing them to cast interpretations that resonate with their interests and ways of perceiving. For example, describers are trained to say that someone is crying, letting AD users conclude whether this suggests the person is upset or not (Snyder, 2005, p. 196). Yet, describers must still prevent listeners from drawing interpretations that are too far distanced from the inferences made by fellow audiences, as this risks hampering AD users' participation in collective responses to media content. Thus, an important responsibility facing audio describers is to pave the way for an expanded variety of interpretations while limiting these possibilities with certain parameters.

Describers are extremely conscious of how they could be impacting listeners' encounters. This heightened self-awareness has partly sought to satisfy injunctions to maintain objectivity (Snyder, 2005, pp. 195–196), but it equally contributes towards burgeoning efforts to harness describers' subjective experiences when enriching descriptions (Mills, 2015; Thompson, 2018). While there are national guidelines that describers may use to inform decisions about what to verbalise, they are in charge of adapting said guidelines to suit the various genres of content (Fryer, 2016, p. 166), and the need for self-reliance is heightened through the limited interaction with artists and their intentions. Hence, audio describers increasingly leverage their own affective and somatic responses as entry points through which to better comprehend, and therefore represent, the content to AD users. For instance, AD training advises the use of multisensory language (Kleege and Wallin, 2015), inviting describers to imagine how something they can see, such as colour, may be represented through another sensory modality and still carry the same or similar meanings. Describers may draw on their own past experiences here, with some even performing the movements they need to describe to develop tactile appreciation and terminology for these features (Balasubramanian, 2019). When it comes to script delivery, describers are aware that their language, timbre, and intonations are heard alongside the auditory qualities of any audiovisual content, making it necessary for them to tailor their speech to suit and reinforce the style of the material being described. In the course of scoping possibilities for how AD users may interpret, describers are tasked with intensely reflecting on and creatively mobilising their own understandings and embodied experiences of the media content.

Acknowledging the creativity that goes into AD production has propelled its status as an art form (Thompson, 2015) and bolstered attempts to regard access overall as a creative tool (Porter, 2016; Romero-Fresco, 2021). This acceptance is evidenced in the inclusion of describers as collaborators early on in the development process. Directors can work with describers as much as they would with another artist in the team, such as a costume designer in theatre (Udo and Fels, 2010, p. 217). Such movements towards foregrounding AD have supported the rise of integrated AD, a specific format which rejects the tendency to incorporate description as a retrofit access measure, and instead, treats it as a core part of the storytelling that contributes to the quality of the overall content (Fryer, 2018, p. 177). The integration of AD in this way has altered production workflows, giving rise to approaches like accessible filmmaking, which intertwines attention to access with the filmmaking process (Romero-Fresco, 2019). Now it becomes possible to explore how auditory features, such as sound design, can elevate and be elevated by the quality of AD (Lopez et al., 2020; Hutchinson and Eardley, 2023).

By valuing artistry and subjective experience as fundamental to AD practice, it becomes possible to endorse a broader range of description styles and describers. A strengthened preference for multisensory language has invited recognition of how art can be appreciated from diverse, often non-normative positionalities, including blind and partially blind audiences. This has opened pathways for co-created AD, whereby descriptions are composed collectively in groups of blind and non-blind people through a question-and-answer method that ensures everyone participates in interrogating and developing the AD. Alison Eardley and colleagues suggest that colour, for example, could be represented through the objects and sensations associated with them, such as a "sandy yellow", "lavender blue", "dusty grey", "terra cotta", and "warm and earthy shades" (Eardley et al., 2022, p. 159). As soon as the colour is identified, the exercise of getting to know the colour and its role in the art encounter is distributed across the co-creators and the personal sensory experiences that may be shaping their attention. Far from being a service offered by "sighted" people to aid blind and partially blind audiences, AD emerges as a means for everyone to explore artwork, no matter their abilities. Authority is shared with untrained audio describers, namely lay audience members, and improvisatory descriptions are legitimised. This flexibility about who can describe and how has arguably accelerated AD's adoption across a number of different contexts, an uptake I seek to expand.

In the following sections, I build on arguments about AD's creative and critical characteristics to propose deepening its integration in design practice. The benefits of incorporating AD for blind and partially blind designers have already been demonstrated. For example, Josh Miele, a Blind designer, drew on participatory descriptions, as well as Touch Graphics, to create tactile and audio maps of the Bay Area Rapid Transit system (Miele et al., 2006). While this remains an important reason for my stance, I will make a case for valuing AD that goes beyond its contribution to non-visual access. I suggest that AD can support designers in their responsibilities to reflect, make futures, and collaborate with user communities. In the process of adapting a disability arts practice for design, it is vital to ask what this means for considerations of justice, especially the interests of disability communities whose contributions have often been appropriated (Jackson, 2018). How can we ensure that disabled people continue to hold a claim to AD when integrated into design and can use it as a channel for launching disability-led interventions (Brilmyer and Lee, 2023)

Co-produced Reflexivity

Reflexivity has been slow to emerge in design practice, and where it is deployed, it tends to be conducted on the terms set by the designers rather than those that are co-produced with user communities. The anxieties about reflecting on designers' positionality (see Fry et al., 2015) are rooted in pragmatist, cognitivist perspectives, which portray designers as experts implementing objective, and hence universal, solutions to problems (Rosner, 2018, pp. 26–30). Surfacing personal experience seems to detract from this approach to design, in which even social and political phenomena are defined with mathematical precision (Rosner, 2018, pp. 31–32). A seminal criticism of this view has been put forward by the design theorist, Donald Schön, who warns that neutralising design risks overlooking the influential role of individuals' experiences and emotions in the formation of knowledge (Schön, 1983). Reflection amongst designers, Schön suggests, is vital for tracing the possible impact of such factors.

Crucially, Schön does not call for a firm departure from a pragmatist lineage or challenge the focus on the individual designer who is apparently uniquely equipped with the tools to reflect. While Schön's intervention helps to start exploring the social and political ways in which individual designers may be positioned (Rosner, 2018, p. 38), it arguably does not urge interrogation of the explanatory concepts and mechanisms that designers may be using to reflect, including how these may be inflected by the material structures, discursive practices, and power relations within which the design process unfolds. The engagement of critical theory may be particularly valuable in this respect (Margolin, 1991). This casts aside a reluctance towards critical theory in design practice and education, a reluctance which appears to stem from the long-term perception of design as vocational training since its formalisation during the Industrial Revolution (Thackara, 1988). Potentially one of the most disruptive critical perspectives in design practice, and one that is especially pertinent to this chapter, is the adoption of the feminist notion of "situated knowledge" posited by Donna Haraway (1991). Derived in a research practice context, the notion of "situated knowledge" proposes that a researcher's observations, as well as analysis of these observations, are shaped by their location, former experiences, and standpoint in power structures (Agid, 2012, p. 34). One's situation in specific structures of race and gender, for example, informs the "partial, locatable, critical knowledges" that mediate their world views (Haraway, 1991, p. 191), which prevent the formation of totalising meanings but still make partial meanings possible. Haraway mobilised the figure of the cyborg, an interference with the human category by mechanical elements, to activate imagination of how normalised knowledge formations could be unsettled by practices and commitments from marginalised communities (Rosner, 2018, p. 55).

Some of the burgeoning critiques of design in the lead-up to the 1990s accelerated the adoption of Haraway's contribution in design practice (Rosner, 2018, p. 42). Lucy Suchman's engagement with Haraway's theory and her conception of "located accountability" (Suchman, 2002) allows consideration of the relationships that influence how design knowledge is developed, particularly facilitating the interrogation of the often-proposed divides between users and designers, as well as boundaries between designers themselves. Going beyond the designer/user opposition, Suchman's theory calls for attention to be given to the complex differences, or multiple situated knowledges, that may come to define our relationships in the design process, and hence, it demands revised conceptions of working relations in design that capture the movement between these differentiated positions (Agid, 2012, p. 34). Suchman helps to recognise how design is a processual way of knowing, a position which is articulated in Design Anthropology (Gunn et al., 2013). Here, understanding is not implemented in design, but it develops out of the context in which design takes place. There is a need to avoid transplanting the ready-made methods for reflection that are popular in much practitioner-targeted literature (Akama and Prendiville, 2013) and prepare for the dynamic ways in which differences can be expected to surface during the design process. The likelihood of unanticipated differences has been amplified by increasingly interdisciplinary collaborations (Furniss, 2020). A realisation that design involves a level of enquiry has encouraged greater interaction with prospective users and communities who may be otherwise affected by design outcomes. These shifts introduce a plethora of empirical experience and multiple ontologies, which will inform how designers reflect. Although existing scholarship has proffered routes towards critical self-reflection (Prakash, 2022), the compulsion to engage with so many diverse perspectives early on in the design process appears to require a means of co-reflecting. Participatory reflexivity may also open avenues for co-creating understandings about the relationships in design, such as the relationships between users and designers. This could further disrupt traditional hierarchies, allowing communities to have more control over the terms on which they participate and undermining extractive knowledge gathering (Costanza-Chock, 2020).

I contend that AD offers a method for co-producing reflexivity, an assertion that I have founded on previous revelations about AD's affordances for reflection in collaborative arts education (Kleege and Wallin, 2015; Thomson, 2020). Kleege and Wallin (2015) suggest exercises inspired by AD, including a participatory technique for students to co-reflect on each other's interpretations of a photograph. By verbalising the described object in terms of the experiences and references they are drawing on to make sense of it, students offer each other multiple routes through which their understandings of these objects could be formed. Expanding these authors' analysis, I want to attend to how the process of describing for others involves considering what listeners need to hear about the source content that can add to their existing encounter with it, as well as what descriptions might make sense to them given previous encounters. As AD scholars have insisted, description also incurs intense self-awareness, whereby describers examine how their relationship with the content might be mediating what they say. Therefore, when composing descriptions, describers are reflexively examining their conceptions of how they, their listeners, and the source content are intersubjectively related. When AD is co-created, a describer's implied perceptions of these relationships, especially of how people are differentially located, are continually intercepted and moulded by other describers. This is how the reflexivity resulting from AD can be co-produced.

In design practice, regular attempts to co-create AD of the content at various stages of development will enable collaborators to assert ideas about how they may be differentially located throughout. Unlike the photograph in Kleege and Wallin's example, designers who describe will not always be looking at a finished piece of work. Instead, they may be verbalising features of the intended shots as these are selected and arranged for filming, such as the landscape, people, or angle. With AD being introduced throughout their creative process, collaborators' descriptions, and anticipations of what each other might find interesting about the content in this moment will be sites for exposing how their intentions for the final outcome may vary. Rather than reflecting based on predefined categories of difference, it will be possible to co-produce a reflexivity that is tailored to the emergent notions of difference in the design process. A model for reflexivity that considers difference to be emergent is imperative, especially for disabled collaborators. Disability communities rarely, if ever, have opportunities to define how they are differentially situated in the design process, as they are frequently divided into specific user groups based on premade, medicalised labels (Hofmann et al., 2020). Advocates for accessibility note how their lived experience of disability is not a constant but can transform as people's relationships with their environment alter (Hamraie and Fritsch, 2019). A further consequence of a flexible notion of difference is that this can permit the statement of intersectional identities (Crenshaw, 1991), which refuse to fit neatly into anticipated categories.

Harnessing this disruptive potential is vital if designers want to heed warnings that reflexivity could become a mere validating mechanism that reinforces hegemonies (Lynch, 2000).

Situated Future-Making

This section extends the above argument about AD's value in uncovering how design knowledge is intricately situated. I illustrate how this application of AD can be harnessed with regard to an activity that is central to all design practice, namely, future-making. The intrinsic futurity in design becomes apparent when acknowledging that creations are intended to transform certain existing aspects of life (Mazé, 2017), often by claiming to fix a problem and articulating preferred, imagined possibilities for enriching human experience. However, rigid recipes for perfect futures are constantly vulnerable to revision by the global and local forces influencing markets, policies, cultural trends, and resource availability. Far from constructing a specific, stable future, therefore, designers need to consider the potential for whatever they design to become something else: be used in unintended ways, take on different meanings, and be mediated through various modalities. The anthropologist, Sarah Pink, and her co-authors (Pink et al., 2018) posit an instructive conception of the future as possibilities, helping to capture the ever-changing world in which designers' creations become a part. As opposed to professing or instigating complete futures for users to inhabit, it appears design contains openings for users to exploratively co-create design outcomes (Smith et al., 2016). With this in mind, future-making practices in design are demanded to explore and shape how diverse possibilities could emerge, rather than specify particular possibilities. Gauging how something could exceed what it superficially seems to be begins with the recognition that users' perceptions are unique to their context and cannot be fully anticipated by designers whose predictions are bounded by how their knowledge is situated. The possibilities that designers are able to make way for may be further shaped by their conceptual frameworks for future-making, including ontological assumptions about connections between past, present, and future (Grosz, 1999, p. 18). As Mazé (2017, p. 38) argues, designers' ideas about what should be different and what should stay the same are not neutral statements but pose political questions with specific consequences and promote particular world views. Critical reflection that pays particular attention to how future-making activities are situated is arguably necessary in design practice. I suggest that co-created AD can document the possibilities that may emerge at different stages of the design process, in addition to disclosing how these possibilities are being shaped by the relationships that collaborators understand themselves to be occupying.

Description does not seek to represent a pre-existing reality but outlines different ways in which phenomena may become experienceable (Marcus et al., 2016). In other words, descriptive practices, like AD, are future-oriented. It is worth recalling, for example, how AD training recommends describers make way for multimodal routes to appreciating content. Here, the presence of dark skies may not just be sensed through verbalising the level of light, as evoking their sombreness, for instance, may conjure up the emotions through which some listeners can infer meaning. What is more, in their attempt to enable diverse possible ways for experiencing the content, describers are not simply attuned to visual details.

Fryer (2016, p. 31) explains that it may be important to describe sonic features too, as some AD users may identify as deaf/Deaf, and some non-verbal sounds may not be self-explanatory. Describers are also encouraged to be responsive to how certain sensory cues can automatically trigger associations for listeners, such as how hearing a Christmas carol may remind them about the aroma of pine needles, the flavour of mince pies, and the dazzle of fairy lights (Ferrington, 1994). Through employing similes and metaphors, describers create analogous relationships between experiences that may not be addressed within primary intentions for the content but may prove meaningful for listeners in unanticipated ways.

Before investigating the implications for design practice, it should be noted that AD practice is not only concerned with generating possibilities but is also about limiting these possibilities. Restricting the scope of interpretations that AD users can make about the source content is significant for ensuring that they do not fall out of synchrony with fellow audiences. Setting the boundaries of this scope is done with descriptions that provide information that act as framing devices within which listeners can form inferences. An example is offered by Fryer (2016, pp. 34–35) in her advice for describing a hypothetical classroom scene. The audio of moving furniture and characters' words might not always be enough to help listeners infer that the setting is a classroom. Instead, describers are advised to pre-empt the image of a classroom featuring pupils who are fidgeting and their teacher, which provides the frame of reference that listeners can use to understand the sound of scraping furniture and the teacher's words. The limitation of AD users' possible inferences also takes into account describers' reflections on the intersubjective relationships between themselves, the content, and their listeners. For instance, if describers wish for listeners to grasp the minimalist style of a theatre set, they may infuse their descriptions with minimalist language, constraining the level of detail that goes into framing AD users' possible interpretations.

Much like audio describers, designers are scoping the possible futures that their creations may come to enable. As multisensory language involves unpacking the various material manifestations their ideas could assume, AD arguably supplements the experimental and exploratory labour that characterises a major part of design practice. I propose that embedding AD in the design process will allow designers to collaboratively explore the multiple possible experiences that could be afforded by their creation at each stage of its making. By unlocking channels through which their collaborators can appreciate the content, each describer is again engaging in co-producing knowledge about how each other may be making sense of the descriptions and hence how they may be situated. Specifically, designers who describe can reveal how they are each scoping possible futures and how this is shaped by their understanding of the socio-material relationships in which they work.

For blind and partially blind designers, the verbalisation of emergent possibilities during the design process encourages their participation in future-making. This move to make knowledge production accessible to disabled designers is indispensable if practitioners wish to avoid addressing disability in tokenistic ways and start contesting normative futures (Minich, 2016). Centring a practice developed within disability communities will arguably embolden disabled designers to foreground their lived experience in practice and fast-track the consideration of making adaptations for diverse audiences, too. By regarding AD as a critical method, designers could pave the way to similarly reconsidering other current access provisions, treating them less like solutions that uphold normativity when it breaks down and more like disruptions that can rethink dominant approaches.

Problematising the Voice

A third potential contribution of AD practice is arguably the critical sensitivity that it helps to cultivate towards the use of voice in design. Vocal acts, of the verbal and non-verbal kind, are increasingly featured in the design process, and yet, there are few approaches for interrogating what voicing comes to mean in the diverse contexts in which designers operate.

The traction gained by the voice can be attributed to a disruption of ocularcentric hierarchies (Drever, 2019, p. 4), with greater importance being assigned to listening (Schafer, 1993). Aural methodologies have been fostered to aid design problem-solving: these typically involve describing and explaining actions as they are carried out through concurrent verbalisation (Lloyd et al., 1995) or articulating practice after the fact through reflective verbalisation (Wetzstein and Hacker, 2004). Sound design scholars have proposed non-verbal tools for conveying sonic concepts, but these methods continue to suggest the value of the human voice as a means for materialising abstractions. For instance, vocal sketching encourages designers to express their ideas through natural vocal sounds, and these are rendered into interactable synthetic sound objects (Delle Monache et al., 2018). Even designers' encounters with language, practices, and notions that seem to critique the design process have been identified as "back-*talk*" (Tonkinwise, 2007, p. 3). This clear tendency to favour vocality when externalising thoughts is rooted in a deep-seated Euro-American philosophical tradition traceable back to Rousseau and Locke (Weidman, 2014, p. 39) that regards the human voice as a vehicle for the modern subject imbued with associations of rationality, agency, and authenticity. Oliver Sacks, in his exploration of information processing, contends that speech is embodied and is "…an uttering-forth of one's whole meaning with one's whole being – the understanding of which involves infinitely more than mere word-recognition" (Sacks, 1998, p. 85). Sacks calls attention to how speech is more than language and how the voice's material and affective qualities, as well as their sociocultural associations, warrant analysis. Colloquial expressions, like "finding my voice", "giving voice to a community", and "breaking silence", perpetuate the perception of spoken language as constituting political power. Such rhetoric has pervaded attempts to include end-users in the design process (Kanstrup and Bertelsen, 2016), often categorised as participatory design (Spinuzzi, 2005). Through collaborative making and thinking, designers and communities gain opportunities to exchange thoughts about articulating the design problem, understanding what is at stake, and any approaches that might be useful. Dialogical forms of listening, such as narrative enquiry (Jones, 2016), have been promoted as ways of embracing marginalised perspectives to help build design scenarios catering to the complexity and variety of prospective users' needs. In this vein, Barcham (2021, p. 1) has sought to integrate storytelling as an indigenous design methodology, which "can help us to create a space of mutuality that can open up new ways of being in the world".

A preoccupation with voicing as an avenue towards social justice risks sidelining instances where vocalisation is not possible or equitable. This risks misconstruing contributions from disabled people who communicate non-verbally or through support workers as somehow lacking agency (Mitchell et al., 2009). Interrogating what voice signifies for the users trying to speak up in the design process appears to be vital. Failure to examine voice contextually could reinforce a prominent concern about participatory design, namely that the approach engages users on terms set by the designers and hence maintains the hierarchies supposedly subverted by participation. Reluctance to be guided by users' knowledge and experiences risks rendering participation into a form of extraction, intended to inform products and services that are ultimately sold back to consumers with little regard for maximising benefits for them (Costanza-Chock, 2020).

Audio describers hone their capacity to critically explore the voice by first becoming acquainted with its potential. AD scholars are aware that the human voice is a conduit for emotion as much as it is for words (Fryer and Freeman, 2014). Fryer (2010, pp. 205–206) suggests that practising AD makes describers highly sensitive to the complexities of the human voice and how to manipulate these for effect. Features of vocal delivery, including pace, pitch, tone, stress, and segmentation, can transform meaning (Fryer, 2016, p. 90). For example, hesitation may lead listeners to doubt the authenticity of what describers are saying (Fryer, 2016, p. 94). Thus, by learning to craft and leverage the intimacy and authority that the human voice affords, designers who practise AD can become sensitive to the multiple ways in which they, their collaborators, and any user communities they work with may be constructing their speech during the design process.

Describers, and arguably designers, need to negotiate any cultural dimensions of the voice. I suggest that the aforementioned association between voicing and political agency, for example, is why the contents of describers' speech seem representative of their personal views, and hence risky, when the subject matter is highly sensitive. During my research in London's West End, audio describers regularly reported feeling self-conscious and even uncomfortable when delivering descriptions about certain topics, including an individual's body, racial and gender identity, sexual content, and violence. Indeed, they might not generally discuss these topics out loud or use certain terminology publicly, particularly from a place of influence; their role as describers compels them to do so. The sense of unease is heightened when describers recall their positionality and question whether it would be appropriate for them to authoritatively describe another person's identity, for instance, especially if that person represents a marginalised community. A describer's voice can make their different positionality audible, since the voice indexes specific subjectivities, which when repeatedly embodied, can become meaningful markers of regional, class-based, and racial identities (Weidman, 2014, p. 40). Notwithstanding these apparently reasonable excuses for describers' anxieties and subsequent temptations to avoid explicitly verbalising some details, AD guidelines strictly prohibit censorship on the grounds of describers' experiences (Rai et al., 2010, p. 77). Studies amongst blind audiences have noted that information on possibly controversial themes is vital, even if not relevant to the plot, as it means listeners can appreciate artistic choices and partake in any related discussions with fellow audiences (Gerber, 2007). The voice here becomes a site for gauging the intersubjective relationship being sewn between the describer and the content described, as well as viscerally sensing the sociocultural tensions generated.

The carefulness around how describers should speak reveals the role of describers' voices in sensing and managing the wider network of intersubjective relationships within which AD users' understanding unfolds. A primary consideration here is how AD is received alongside other sonic elements within the broader experience of the media content. When AD is imbedded in dialogue and is heard through the same speaker output, like in a home screening of a film, it may not be possible for visually impaired audiences to filter out the AD of distressing images, such as during depictions of violence, whereas audiences seeing the content may be able to temporarily glance away or close their eyes (Fryer, 2016, p. 148). Meanwhile, AD at live events may be heard through headphones, meaning that listeners receive potentially disturbing descriptions directly in their ears and experience an intimacy with material that may not be welcome. Describers find they must also adjust their speech to institutional constraints. Venues closely monitor and police against descriptions of bodies for the sake of political correctness (Fryer, 2018, p. 181). What is determined to be politically correct can vary constantly (Fryer, 2016, p. 152), making it necessary for describers to somehow select terminology that could be suitable across time and cultural contexts. In the process of verbalising content that they find uncomfortable, describers attune to what their voice, and vocality in general, can do and come to signify in a given context. This attuning could arguably be identified as a listening practice, one which is deeply embodied rather than only involving audition.

I argue that designers can effectively harness the ways in which AD helps to listen to the voice as part of critically deploying verbal acts in design. Hence, I build on scholarly precedents that distinguish listening from hearing, specifying the latter as a passive absorption of auditory stimuli dependent on the physiological ear (Collins, 2020, p. 9). On the other hand, listening has emerged as a mode of attention. The theorist, Brandon LaBelle (2020, p. 563), posits listening as an embodied, active, multimodal practice that creates opportunities for enquiry, as well as relationships and imaginings. According to LaBelle, attending to acoustics, that is, how sound has been conditioned or enabled to move, permits examination of how the articulation of views and those expressing them are managed (LaBelle, 2020, p. 550). Consequently, listening involves becoming sensitive to how "one synchronises, attunes, and aligns with others by way of what one hears and feels, and how bodily or affective experiences support forms of participation" (LaBelle, 2020, p. 552). I seek to supplement LaBelle's departure from narrow conceptions of listening with the assertion that attending to the affective and somatic frictions encountered when speaking is a form of listening. This would allow designers to listen to the tensions that emerge through their own voice when they speak, as well as acknowledge and value the challenges faced by any user communities who try to speak as required in participatory design projects.

Manifesto

The following five principles capture the intended values behind integrating AD in design practice. I only offer loose guidelines for realising these values in practice, wanting to leave room for adaptation to designers' specific circumstances.

Improvisatory – Designers can describe any aspect of the design, whether it is deemed visual or not. These descriptions can be composed live, rather than being scripted or pre-recorded. This encourages descriptions to be reflective of how describers are situated in time, place, and stage of the design process.

Cross-modal – AD practice believes that words can capture the meaning in non-verbal phenomena. Designers should extend this cross-modality in their descriptions by imagining how experiences of a given phenomenon could be mediated through different modalities and then articulating these possibilities with multisensory language. Examples include describing the sound of a blaring horn through references to an angry red colour, searing heat, or a sour fruit. Tapping into various bodily experiences through the descriptions, particularly without referencing the original cause of the sound, helps to quickly get past accepted ways of describing and urges participants to foreground their subjectivities faster for reflection.

Speculative – AD and design are both future-making practices. By anticipating how listeners experience the content, describers expose multiple possibilities that reflect their perceptions of the intersubjective relationships in which they are describing and designing. The emphasis is not on formulating stable futures but on using AD to understand how these possibilities are crafted.

Participatory – Maximising the critical and reflexive benefits of AD is possible through co-creating descriptions with and for a diverse selection of positionalities throughout the design process, including collaborators and potential end-users. Everyone involved in design may describe their experiences of the content as it stands.

Vocal – Designers should try to exchange descriptions out loud with collaborators, tracing the frictions encountered by all while doing so. This problematises the epistemological privileging of "voice" in design, as well as urging analysis of voices' material and symbolic qualities in situ. In AD practice, pauses define speaking as much as verbal expression. Similarly, designers should pay attention to both audible vocals and silence when studying the voice. Appreciating the potential of the voice helps designers to manipulate its usage and recognise how it is being leveraged by their collaborators.

Conclusion

To close, I contemplate how the manifesto may be applied, an evaluative exercise that casts my contribution in this chapter as the humble start to a much larger project. Through demonstrating how AD critically enriches currently prioritised dimensions of design, namely reflection, future-making, and voicing, I have proposed potential routes through which AD practice can be incorporated in the education of upcoming designers. As a methodology for articulating and collaboratively negotiating difference, AD can support designers to navigate the increasingly interdisciplinary contexts where they work, as well as invite marginalised communities to critically interrogate and redefine their engagement in the design process. It should be noted that the AD practices addressed here have been developed in a Euro-American context where the UK and the USA have held the most influence as the first test beds for AD in the 1990s, with creative and pedagogical input following later from wider Europe. While outside the scope of this present chapter, there is strong scholarship discussing AD norms in other regions (Bittner, 2012; Leung, 2015), which designers working in those regions need to consider when adapting this manifesto.

AD proves particularly relevant in design contexts where audio is a key category that warrants reflection, such as sound design. Whether sharing early proofs of concept or developed prototypes amongst team members, designers may attempt to describe what a sample or track sounds, or even feels, looks, and smells like, to name a few suggestions. Partaking in these improvised, cross-modal descriptions enables designers to co-produce knowledge of what is being designed in a way that exposes and interrogates the social relations constituting the sound design process. For sound designers, who are conventionally deprioritised in interdisciplinary teams, AD presents an opportunity to discuss and reshape these differentiations in unprescribed ways. When speculating on how audio can be experienced variously by other team members, AD becomes an aural methodology for tracing how sound designers make multiple possibilities emerge and the role of audio in informing how they imagine otherwise. Any abstract insight garnered about participation and vocality can be rapidly materialised, iteratively developed, and hence critiqued through the auditory practices that sound designers have at their disposal.

A responsibility facing designers who wish to integrate AD in their practice is to ensure that there remains scope for disability-led intervention. The valuing of access practices as important skills could arguably catalyse engagement with them, even when not creating output for disabled audiences, and hence find recognition for the expertise held in disability communities. This is intended to open doors for more disabled designers, such as blind and partially blind people, who can work with the auditory modality preferred by AD. Indeed, the auditory preference in AD may become a restriction for widening adoption, and this is why the above manifesto is incomplete without more manifestos advocating for the integration of multiple other access practices. The deliberate emphasis on embodied listening in AD opens pathways for these other access practices to come through, a justice afforded at sound's boundaries.

References

Agid, S., 2012. World making: Working through theory/practice in design. *Design and Culture*, 4 (1), pp. 27–54.

Akama, Y. and Prendiville, A., 2013. Embodying, enacting and entangling design: A phenomenological view to co-designing services. *Swedish Design Research Journal*, 1, pp. 13–29.
Balasubramanian, H., 2017. Thinking about vision: An investigation with visually-impaired theatre-goers and audio describers. *Camthropod*, [online] Available at: https://soundcloud.com/sianml/episode-14-thinking-about-vision-by-harsha-balasubramanian [Accessed 14 March 2022].
Balasubramanian, H., 2019. Experiencing writing: Lessons for multimodal ethnographers from audio describers of dance. *Entanglements*, 2 (1), pp. 188–201.
Barcham, M., 2021. Towards a radically inclusive design–indigenous story-telling as codesign methodology. *CoDesign*, 19 (1), pp. 1–13.
Bittner, H., 2012. Audio description guidelines: A comparison. *New Perspectives in Translation*, 20, pp. 41–61.
Braun, S., 2008. Audio description research: state of the art and beyond. *Translation Studies in the New Millennium*, 6, pp. 14–30.
Brilmyer, G. and Lee, C., 2023. Terms of use: Crip legibility in information systems. *First Monday*, [online] Available at: https://firstmonday.org/ojs/index.php/fm/article/view/12935 [Accessed 5 January 2024].
Chottin, M. and Thompson, H., 2021. Blindness gain "as worldmaking: Audio Description as a new" partage du sensible. *L'esprit créateur*, 61 (4), pp. 32–44.
Collins, K., 2020. *Studying Sound: A Theory and Practice of Sound Design.* Cambridge: MIT Press.
Costanza-Chock, S., 2020. *Design Justice: Community-led Practices to Build the Worlds We Need.* Cambridge, MA: The MIT Press.
Crenshaw, K., 1991. Mapping the margins: Intersectionality, identity politics, and violence against women of colour. *Stanford Law Review*, 43 (6), pp. 1241–1299.
Delle Monache, S., Rocchesso, D., Bevilacqua, F., Lemaitre, G., Baldan, S. and Cera, A., 2018. Embodied sound design. *International Journal of Human-Computer Studies*, 118, pp. 47–59.
Di Giovanni, E., 2018. Audio description for live performances and audience participation. *The Journal of Specialised Translation*, 29, pp. 189–211.
Drever, J.L., 2019. 'Primacy of the Ear'–But Whose Ear?: The case for aural diversity in sonic arts practice and discourse. *Organised Sound*, 24 (1), pp. 85–95.
Eardley, A.F., Thompson, H., Fineman, A., Hutchinson, R., Bywood, L. and Cock, M., 2022. Devisualizing the museum: From access to inclusion. *Journal of Museum Education*, 47 (2), pp. 150–165.
Ferrington, G., 1994. Audio design: Creating multi-sensory images for the mind. *Journal of Visual Literacy*, 14 (1), pp. 61–67.
Fryer, L.E., 2010. Audio description as audio drama–a practitioner's point of view. *Perspectives: Studies in Translatology*, 18 (3), pp. 205–213.
Fryer, L. E., 2016. *An Introduction to Audio Description: A Practical Guide.* London: Routledge.
Fryer, L.E., 2018. The independent audio describer is dead: Long live audio description! *Journal of Audiovisual Translation*, 1 (1), pp. 170–186.
Fryer, L.E. and Freeman, J., 2014. Can you feel what I'm saying? The impact of verbal information on emotion elicitation and presence in people with a visual impairment. In: A. Felnhofer and O.D. Kothgassner, eds. *Challenging Presence: Proceedings of the 15th International Conference on Presence.* Vienna: facultas. wuv, pp. 99–107.
Furniss, L., 2020. Beyond discipline: Evolving design practice and design education in the twenty-first century. *Architecture_MPS*, 18 (1), pp. 1–17.

Gerber, E., 2007. Seeing isn't believing: blindness, race, and cultural literacy. *The Senses and Society*, 2 (1), pp. 27–40.

Grosz, E., 1999. *Becomings*. Ithaca: Cornell University Press.

Gunn, W., Otto, T. and Smith, R.C., 2013. *Design Anthropology: Theory and Practice*. London, New York: Bloomsbury.

Hamraie, A. and Fritsch, K., 2019. Crip technoscience manifesto. *Catalyst: Feminism, Theory, Technoscience*, 5 (1), pp. 1–33.

Haraway, D.J., 1991. *Simians, Cyborgs, and Women: The Reinvention of Nature*. New York: Routledge.

Hofmann, M., Kasnitz, D., Mankoff, J., and Bennett, C. L., 2020, October. Living disability theory: Reflections on access, research, and design. In *The 22nd International ACM SIGACCESS Conference on Computers and Accessibility*, pp. 1–13. Association for Computing Machinery.

Hughes, C., Orero, P. and Rai, S., 2020. Towards a user specification for immersive audio description. In: S. Braun and K. Starr, eds. *Innovation in Audio Description Research*. London: Routledge, pp. 121–134.

Hutchinson, R. and Eardley, A.F., 2023. 'I felt I was right there with them': The impact of sound-enriched audio description on experiencing and remembering artworks, for blind and sighted museum audiences. *Museum Management and Curatorship*, 39 (6), pp. 733–750.

Hutchinson, R.S. and Eardley, A.F., 2020. The accessible museum: Towards an understanding of international audio description practices in museums. *Journal of Visual Impairment & Blindness*, 114 (6), pp. 475–487.

Ibáñez, A.M. and Vermeulen, A., 2013. Audio description as a tool to improve lexical and phraseological competence in foreign language learning. In: D. Tsagari and G. Floros, eds. *Translation in Language Teaching and Assessment*. Newcastle upon Tyne: Cambridge Scholars Publishing, pp. 45–61.

Jackson, L., 2018. We are the original lifehackers. *New York Times*, [online] Available at: https://www.nytimes.com/2018/05/30/opinion/disability-design-lifehacks.html [Accessed 14 March 2022].

Jones, N.N., 2016. Narrative inquiry in human-centered design: Examining silence and voice to promote social justice in design scenarios. *Journal of Technical Writing and Communication*, 46 (4), pp. 471–492.

Kanstrup, A.M. and Bertelsen, P., 2016, August. Bringing new voices to design of exercise technology: Participatory design with vulnerable young adults. In *Proceedings of the 14th Participatory Design Conference: Full Papers-Volume 1*, pp. 121–130. Arhus, Denmark: Association for Computing Machinery.

Kleege, G. and Wallin, S., 2015. Audio description as a pedagogical tool. *Disability Studies Quarterly*, 35 (2), [online] Available at: https://dsq-sds.org/index.php/dsq/article/view/4622/3945 [Accessed 15 March 2022].

LaBelle, B., 2020. Towards acoustic justice. *Law Text Culture*, 24, pp. 550–572.

Leung, D., 2015. Audio description in Hong Kong. In: R.B. Piñero and G.D. Cintas, eds. *Audiovisual Translation in a Global Context*. London: Palgrave Macmillan, pp. 266–281.

Lloyd, P., Lawson, B. and Scott, P., 1995. Can concurrent verbalization reveal design cognition? *Design Studies*, 16 (2), pp. 237–259.

Lopez, M., Kearney, G., Hofstädter, K. and Balla, G., 2020. Enhancing audio description: Accessible filmmaking, sound design and the importance of educating filmmakers. *Media Practice and Education*, 21 (4), pp. 289–304.

Lynch, M., 2000. Against reflexivity as an academic virtue and source of privileged knowledge. *Theory, Culture & Society*, 17 (3), pp. 26–54.

Marcus, S., Love, H. and Best, S., 2016. Building a better description. *Representations*, 135 (1), pp. 1–21.

Margolin, V., 1991. Design studies and the education of designers. *ELISAVA TdD*, 6, [online] Available at: http://tdd.elisava.net/coleccion/6/margolin-ca [Accessed 15 October 2021].

Mazur, I. and Chmiel, A., 2021. Audio description training: A snapshot of the current practices. *The Interpreter and Translator Trainer*, 15 (1), pp. 51–65.

Mazé, R., 2017. Design and the Future: Temporal Politics of 'Making a Difference'. In: R. Smith, K. Vangkilde, M. Kjersgaard, T. Otto, J. Halse, and T. Binder, eds. *Design Anthropological Futures*, London: Bloomsbury, pp. 37–54.

Miele, J.A., Landau, S. and Gilden, D., 2006. Talking TMAP: Automated generation of audio-tactile maps using Smith-Kettlewell's TMAP software. *British Journal of Visual Impairment*, 24 (2), pp. 93–100.

Mills, M., 2015. Listening to images: Audio description, the translation overlay, and image retrieval. *The Cine-Files*, 8 (Spring Issue), [online] Available at: https://www.thecine-files.com/listening-to-images-audio-description-the-translation-overlay-and-image-retrieval/ [Accessed 15 October 2021].

Minich, J.A., 2016. Enabling whom? Critical disability studies now. *Lateral*, 5 (1), [online] Available at: https://www.jstor.org/stable/48671434 [Accessed 7 September 2023].

Mitchell, W., Franklin, A., Greco, V. and Bell, M., 2009. Working with children with learning disabilities and/or who communicate non-verbally: Research experiences and their implications for social work education, increased participation and social inclusion. *Social Work Education*, 28 (3), pp. 309–324.

Pihkala, S. and Karasti, H., 2016, August. Reflexive engagement: Enacting reflexivity in design and for 'participation in plural. In *Proceedings of the 14th Participatory Design Conference: Full Papers-Volume 1*, pp. 21–30. Arhus, Denmark: Association for Computing Machinery.

Pink, S., Akama, Y. and Sumartojo, S., 2018. *Uncertainty and Possibility: New Approaches to Future Making in Design Anthropology*. London: Bloomsbury Publishing.

Porter, L., 2016. Audio description: The art of access. *Disability Arts Online*, [online] Available at: https://disabilityarts.online/magazine/opinion/audio-description-art-access/ [Accessed 14 March 2022].

Prakash, S., 2022. Preparing for the pluriverse: Embracing critical self-reflection in service design practice. In: D. Lockton, S. Lenzi, P. Hekkert, A. Oak, J. Sádaba, and P. Lloyd, eds. *DRS2022: Bilbao, 25 June- 3 July*. Spain: Bilbao.

Rai, S., Greening, J. and Petré, L., 2010. *A comparative Study of Audio Description Guidelines Prevalent in Different Countries*. London: RNIB, Media and Culture Department.

Remael, A. and Vercauteren, G., 2007. Audio describing the exposition phase of films: Teaching students what to choose. *Trans: Revista de Traductologia*, 11, pp. 73–93.

Romero-Fresco, P., 2019. *Accessible Filmmaking: Integrating Translation and Accessibility into the Filmmaking Process*. London: Routledge.

Romero-Fresco, P., 2021. Creativity in media accessibility: A political issue. *Cultus: The Journal of Intercultural Mediation and Communication*, 14, pp. 162–197.

Romero-Fresco, P. and Chaume, F., 2022. Creativity in audiovisual translation and media accessibility. *Journal of Specialised Translation*, 33, pp. 75–101.

Rosner, D.K., 2018. *Critical Fabulations: Reworking the Methods and Margins of Design*. Cambridge, MA: The MIT Press.

Sacks, O., 1998. *The Man Who Mistook His Wife for a Hat and Other Clinical Tales*. New York, NY: Touchstone.

Schafer, R.M., 1993. *The Soundscape: Our Sonic Environment and the Tuning of the World*. London: Simon and Schuster.
Schön, D.A., 1983. *The Reflective Practitioner: How Professionals Think in Action*. New York: Basic Books.
Snyder, J., 2005. *The Visual Made Verbal*. International Congress Series 1282, Elsevier, pp. 935–939.
Spinuzzi, C., 2005. The methodology of participatory design. *Technical Communication*, 52 (2), pp. 163–174.
Suchman, L., 2002. Located accountabilities in technology production. *Scandinavian Journal of Information Systems*, 14 (2), pp. 91–105.
Thackara, J., 1988. *Design After Modernism*. New York: Thames and Hudson.
Thompson, H., 2015. How the arts can help change attitudes to blindness. *The Guardian*, [online] Available at: https://www.theguardian.com/culture-professionals-network/2015/aug/03/arts-change-attitude-blindness-blind-creations [Accessed 15 March 2022].
Thompson, H., 2018. Audio description: Turning access to film into cinema art. *Disability Studies Quarterly*, 38 (3), [online] Available at: https://dsq-sds.org/article/view/6487/5085 [Accessed 10 September 2021].
Thomson, E.A., 2020. Audio description as a collaborative and reflexive tool (Practice Brief). *Journal of Postsecondary Education and Disability*, 33 (1), pp. 103–107.
Tonkinwise, C., 2007. My theory practice joys. *Lecture presented at the Graduate Research Conference, School of Design and Architecture*, RMIT University, Melbourne, Australia, 19–21 October 2007.
Udo, J.P. and Fels, D.I., 2010. The rogue poster-children of universal design: Closed captioning and audio description. *Journal of Engineering Design*, 21 (2–3), pp. 207–221.
Weidman, A., 2014. Anthropology and voice. *Annual Review of Anthropology*, 43 (1), pp. 37–51.
Wetzstein, A. and Hacker, W., 2004. Reflective verbalization improves solutions—the effects of question-based reflection in design problem solving. *Applied Cognitive Psychology*, 18 (2), pp. 145–156.
Williamson, B., 2012. Electric moms and quad drivers: People with disabilities buying, making, and using technology in postwar America. *American Studies*, 52 (1), pp. 5–29.

3

FOR SOUND TO EMANATE FROM SITES

Or, Some Notes on Black Sonic Bildung

Nathanael Kliebhan

Black Sound—Space—Coloniality—Poetics—Afro-German Literature

Introduction

What follows is a tracing of the *perforation* of imposed enclosures that black sonic interventions enable. Without aiming to form a coherent structure composed as a linear progression of sections, this text consists of *passages* of varying scales that are not isolated from each other. With such an assemblage of textual anti-colonial and anti-anti-black passages, it is intended to explore the topos formed by black study, sound, and space.

In the first passage, we look at the idea of *Naturbeherrschung* and engage with Fred Moten and Stefano Harney's critique thereof, enabling us to articulate that logics of enclosure are already reductive, already racial (Harney & Moten, 2013). Then we engage with Frankfurt School's influential (albeit insufficient) critique of *Naturbeherrschung*, against which is employed a statement by DJ and producer Theo Parrish as a case for how black experiments with sound (here techno*) deform* technological implementation against a technological administration of (human) nature (Parrish, 2012). Such a critique of *Naturbeherrschung* will pose the backdrop for the remaining pages and will be a first attempt at tracing the *black sonic.*

In the next passage, "An Anti-Colonial (Sonic) Phenomenology of Spatialities Beyond Outside/Inside", based on an argument that Gaston Bachelard develops in *Poetics of Space*, we pose that the opposition of outside and inside is a formality emerging from Occidental representations of space. Outlining how an unambiguous distinction between outside and inside is the ontological basis for the racial/colonial project we focus on how the coloniality of this

DOI: 10.4324/9781003194453-4

strict geometry is reified in logics of fixed spatial demarcation (property, territory) and, in Sylvia Wynter's terms, in the figuration of the normative Subject as *Man*. Based on a short phenomenological meditation, we will think through how a sonic approach to spatiality unsettles fixed geometric demarcations and derive how this post-dualistic spatiality provides an anti-colonial opening.

"(Non)Performance, Perforation and the Gesture of the Black Sonic (As Disruptive Noise)" outlines the passage of (resistive) black sound through walls of confinement. With a reading of Saidiya Hartman's "Riot and Refrain", which is indispensable for this undertaking, and late May Ayim's "Aufbruch", whose voice remains a guide throughout, we will trace the *perforative* gesture that is at the core of this inquiry. After the preceding more theoretical engagements, we will deviate from the "academic script" and begin to rely on more poetic formulations. While this passage directly engages with the themes of this inquiry, it is not intended to spell out a "lucid" link.

The short passage entitled "Miniatures From Displace Between" is comprised of a poem I wrote and a reading of a quote by May Ayim with which she articulates her experience after the fall of the Berlin Wall. It is a short interlude that will move us into black Germany, which is the terrain of the remaining pages. The final passage, "Black Sociality as Sonic Intimacy?" is a personal offering. From my experience in Riedlingen—my hometown in rural Germany—I narrate my entanglement in the themes that have been established—sonic permeation over geometric spatial demarcations (in the environment of rural Germany) and (the absence of) black sociality. Besides, an implication of this passage is a black sonic unsettling of a strict separation of rural and urban spheres.

Preliminary Remarks

I am not a theorist. My use of theoretical fragments will, for this reason, necessarily evidence a degree of inadequacy or incompleteness. Not seeing this as a fault, I aim to employ this fragmented approach as a method of being in dialogue with thinkers that fascinate me, while not being firmly settled in the "canon".[1] one must not first acquire a deep insight into the philosophical tradition to write from the unsettled position of a black German person or to practice black critique.[2]

Also, *I am not a poet.* Yet, as this research progressed, I began to sense an urge to include poetic and autobiographical fragments as a method of tracing from which dissonant "locus" this research has been undertaken.[3] See this as an experiment. It is fruitful, in my view, to use fissures as potential diversions towards other ways of being in and *with* this world, or—using a phrase by black German feminist pioneer May Ayim—to use this *Zwischenwelt als Chance* [world in between as an opportunity] (Ayim, 2020b). I suggest reading this text as immediately formed by this ambiguity-tolerating terrain, as continual formulation, and as such a basis for further—perhaps, sonic—inquiry.

Striving to speculate on the resistive force of black sound (beyond enclosure, logics of property, self-possession) comes with the risk of performing a tame song of praise, especially if I were to conceal that what is centred is specific to my personal and contingent idea of what black sound is. As a starting point, therefore, I will "define" what the black sonic implies throughout this engagement.

Definition and Deformation

What is *black sound*? According to the "academic script", we need a clear definition, yet if we recall the literal meaning of "definition" as the "setting of boundaries", such a task is impossible. It is precisely this impossibility to define, to limit black sound, its ability to permeate beyond enclosure, that this undertaking attempts to engage with. *Black* is here understood to be the critical impetus, the setting against structures of racial/colonial/heteropatriarchal domination that constitute our socio-economic formations and define and constrict black (femme/queer/pathologised/non-human) life. Instead of aiming to find the origin of this impetus, we will look at the urgent *deformations* that this impetus makes possible. *Sound* is commonly defined as the mechanical *deformation* of a medium that propagates as a wave. Keeping this in mind, we can think about how people across the black diaspora employ sound to deform various structures—structures of language, technology, and even, as we will see, of the built environment. Édouard Glissant, for instance, speaks of dub poetry as the "ingenious, aggressive deformation of one language, the English language…that emerges from the subversive use of that language" (Glissant, 2020). It is this kind of deformation that the current undertaking is animated by, and what we will focus on lies, more specifically, in distilling the gesture of *deformation* and thinking of it concerning the socio-economic spatial formations of racial/colonial modernity.

Black sonic deformation is irreducibly insurgent; it is system-destabilising aesthetic knowledge production and transmission. When approaching any seemingly definite limitation, the force of such deformation can effectively *perforate* it. A closely related element, or perhaps better said, a consequence of *deformation* is, therefore, what we might term *perforation*—the unsettling of unambiguous demarcations between inside and outside—which will remain the central figure of thought throughout this text.

The aim of my remarks is not to adventurously and, perhaps, counterintuitively *abandon* the categories outside/inside. Instead, I wish to draw attention to the fact that beyond Occidental spatial conceptions that picture space set before the eye (*Raumvorstellung*), there are other, indeed sonic, approaches to spatiality (*Raumerfahrung*) that *also* encourage an intuition *beyond* outside/inside. This "also" is crucial: If this essay is an exercise in unsettling unambiguous dualisms, we must also unsettle the dualism between a (visual) approach to spatiality that distinguishes between outside and inside, and a (sonic) approach that doesn't. Sound *has the ability to shield*, to enclose non-violently, to establish intimacy.[4] I want us to focus on perforating black sonic expressions as strategies against the violent *imposition* of enclosure. Here, we should keep in mind Fred Moten's poignant question: Is it "[s]ound that must be heard as if it were in response to that violence, [or] the sound that to which such violence responds?" (Moten, 2018) The way I intend to engage with the *black sonic* emerges from this field of tension; it is certainly not an unambiguous, objective formula. Instead, it is a cypher that is inseparable from what I experience *black sound* to be. This essay is an attempt, and an experiment to echo its boundless complexity, at least metaphorically. Various black sonic practices offer alternative conceptions of space, make walls porous, perforate structures of constriction,[5] *can you not hear the debris falling, the voices, and the music floating?*

Notes on Naturbeherrschung

> I think you are more than another testament of the stark implications of European voyaging. More than an indictment of the rush. More than the folly of a dominant way of living that changes the planet, quickly, thoughtlessly, forever. More than the deadliness of an insatiable hunger born of chasing things other than sustenance. That hunger outlived you. I feel it chasing me too.
>
> *Alexis Pauline Gumbs (Gumbs, 2020, p. 21)*

Naturbeherrschung (German for the "domination of nature") is conventionally used to describe the project of human subjection of nature and its inhabitants through technology and the acquisition of knowledge on natural processes. Understood in this way *Naturbeherschung* is to be read as the maxim of the Enlightenment project and as a guiding principle of the modern world. It fortifies a stable demarcation between the natural and the human sphere, however, we can extend its purview. The term already addresses *Herr* (man or master) which makes *Naturbeherrschung* a gendered term that draws attention to the epistemic figure at the root of the multiple forms of modern subjection.[6] It is a term that is aside from its obvious use as a marker of *Man's* ecological violations, descriptive of the horrors of Atlantic Slavery, and (settler) colonisation because these forms of subjection in their Eurocentric terms did not subject the people along the Gulf of Guinea, in the Americas and beyond as humans, but as part of a "natural" and as such disposable scenery.[7] In other words, the modern socio-economic projects of material redistribution (extraction and extinction), which include, but are not limited to Atlantic slavery, plantocracy, (settler) colonisation, industrialisation, and the sedimentation of modern logics of property, are inseparable from a positive assessment of the idea of *Naturbeherrschung* (Gumbs, 2020).

Naturbherrschung, in its practical implementation in the modern world's racial/colonial formations, naturalises a metaphysics of divisibility.[8] Land is taken, property is enclosed, territory is occupied, and most absurdly, humans are enslaved. Practices of dispossessive possession are where the idea of *Naturbeherrschung* materialises as an eruption of violence that is still rampant. When in early modern Europe land was being enclosed and the modern logic of possession was conceived, "...establishing title to land by making it more productive meant eliminating biodiversity and isolating and breeding a species—barley or rye or pigs" (Harney & Moten, 2021, p. 29). As Moten and Harney remark here, for modern logics of possession spatial division is legitimated by a radically reductive mode of cultivation that is hostile to biodiversity. In this way, the imposition of enclosure is also inherently racialised; they continue:

> [F]rom the outset, the ability to own—and that ability's first derivative, self-possession—is entwined with the ability to make more productive. In order to be improved, to be rendered more productive, land must be violently reduced to its productivity, which is the regulatory diminishment and management of earthly generativity. Speciation is this general reduction of the earth to productivity and submission of the earth to techniques of domination that isolate and enforce particular increases in and accelerations of productivity. In this regard, (necessarily European) man, in and as the exception, imposes speciation upon himself, in an operation that extracts and excepts himself from the earth in order to confirm his supposed domination over it. And just as the earth must be forcefully speciated to be possessed, man must forcefully speciate himself in order to enact this kind of possession. This is to say that racialisation is present in the very idea of domination over the earth; in the very idea and enactment of the exception; in the very nuts and bolts of possession-by-improvement.
>
> *(Harney & Moten, 2021, p. 29)*

In this sense, *Naturbeherrschung* is also domination of human nature, enacted by the figuration of the Self against the ground of its "subhuman" and nonhuman Others, and, therefore, a critique of *Naturbeherrschung* must invariably be a critique of the modern Subject and its racialised, gendered overdetermination (that modern striving towards disambiguation that Moten and Harney term *speciation*).[9]

> Somebody told me it's the end of the world but that's just for some....The Roots, Quicksand Millennium
>
> *(The Roots, 1999)*

A thorough—though, regarding orders of domination such as race and gender fundamentally insufficient—critique of the idea of *Naturbeherrschung* is given by Frankfurt School, most notoriously in Theodor Adorno and Max Horkheimer's *Dialektik der Aufklärun.*[10] They too point to the notion that all human Enlightenment-based strivings to emancipate the human from the natural embedding lead to barbaric domination of nature and of the human itself.[11] However, posing that "[t]oday we are subjected to nature's compulsion and our domination of it is mere conceit" (Horkheimer & Adorno, 1988, p. 10). they prompt that *Naturbeherrschung* is a kind of macabre performance that is not achieved in the emphatic sense And while this might, on the one hand, seem accurate, it is on the other hand dangerously inaccurate because—and this is pronounced in Moten and Harney's quote above—it is specifically *Man's* coercive performance that most violently sustains this conceit at the cost of those who are (and that which is) solely ground to his figure.

Another insufficiency of their critique of *Naturbeherrschung* lies in their claim that all human cultural strivings—already the myth—entail and foster the impetus that would later be the reductionist and totalitarian dogma of modern positivism, entirely neglecting that there are insurgent cultural interventions that are not complicit in but enact the escape from *Man's* domination of nature.[12] A portion of nuance to add to their critique would be to suggest that, if *deformed*, culture, technology—even the machine—is employable against *Naturbeherrschung*.

Faced with the violence and constriction that modern thought exerts on (black) life, black sound practices are a form of cultural production constituting forms of living that bypass the pitfalls of the failed emancipatory project Adorno and Horkheimer describe. In other words, the figure of the self-aggrandising Subject—*Man*—in its attempt to technologically extract itself from nature's bondage, resorts to a *corrosive*,[13] instrumental rationality—the consequence of which is the modern *socioecological disaster*.[14] Within and through black sound practices, on the other hand, the *technological*, and techno-economic formations are *deformed* to bring forth spatial options in which black (social) life is asserted—spaces against corroding rationality, against the getting out of control of the positivist dogma of efficiency and logic, against instrumental reason-driven exploitation that violates Earth and its (non-*Man*) human and non-human inhabitants. Black intervention through sound technology can be a technological implementation beyond instrumental rationality; it asserts cultural life without *Naturbeherrschung*. Detroit offers this in techno. *Are you listening?*

We can corroborate with a short interlocution between sonic pioneer Theo Parrish and Karl Marx, since the Frankfurt School's view that cultural strivings foster an impetus of *Naturbeherrschung* and that, ultimately, industrial society mechanises and assimilates the human is to be traced to a Marxian critique. Walter Benjamin—a close intellectual associate of Adorno and Horkheimer—quotes Marx:

> Aller kapitalistischen Produktion…ist es gemeinsam, daß nicht der Arbeiter die Arbeitsbedingung, sondern umgekehrt die Arbeitsbedingung den Arbeiter anwendet, aber erst mit der Maschinerie erhält diese Verkehrung technisch handgreifliche Wirklichkeit.
>
> *(Benjamin, 1991, p. 631)*

[For all capitalist production…it is a common trait that the worker does not make use of the working conditions, but rather that the working conditions make use of the worker. Though only with machinery this reversal obtains a technologically concrete actuality].

We can respond with DJ and producer Theo Parrish:

> What I think everybody misunderstood, particularly Europe, is that [Techno] is an obsession and an ode to the machine. No! In Detroit, black people are being dominated by the machine. Having to go to work every day to the point where your mind is driven by putting the same screw in the same Cadillac every single day. [Techno is about] breaking out from that but by using the machine in escaping.
>
> *(Parrish, 2012)*

In this short excerpt from an interview with *Telekom Electronic Beats*, Theo Parrish, first confirms and then reverses the Marxian materialistic, socio-economic critique of the machine's harming and assimilating nature. Techno as obsession/valorisation of the machine would be a flat, alienated extension of an aestheticised instrumental rationality; as a fugitive *deformation* of the machine, it reveals the potential of black critique. Black sociality's sonic experimentations invert the machine age application of humans. Here, humans (that were formerly (de)valued as mere commodities/means), while critically reflecting their social reality (ab)use technology on *their* terms (as ends) and in this way enact (and make conceivable) an implementation of technology beyond *corrosive rationality*, beyond *Naturbeherrschung*.

A *stretched* critique of *Naturbeherrschung* in this sense, then, allows us to encircle the Subject, so invested in the subjection of its Others, and to unsettle the unambiguous demarcation between the natural and the human realms.[15] Through black sonic experiments, we *can listen* to a critique of *Naturbeherrschung*, which we can meaningfully mobilise for anti-colonial, anti-racist, anti-ableist, feminist, ecological, and queer interventions. Alexis Pauline Gumbs' guide quote sounds out that the "insatiable hunger born of chasing things other than sustenance" (Gumbs, 2020, p. 21) chased down the Hydrodamalis gigas, whom she is addressing, into its extinction; she can also feel it chasing her as a black queer woman. It, therefore, reads as an enunciation—it would be difficult to exceed its urgency—of our embeddedness in the ensemble of living things, and therefore, the possibility of a formation that transcends logics of speciation, against self-possessing *Man* and his self-legitimising, self-imparted project (Self- and Other destroying) of *Naturbeherrschung*.

With the following quote from Adorno and Horkheimer, we can proceed to the next passage:

> Humans believe themselves free of fear when there is no longer anything unknown....Enlightenment is mythical fear radicalized. The pure immanence of positivism, its ultimate product, is nothing other than a form of universal taboo. *Nothing is allowed to remain outside, since the mere idea of the "outside" is the real source of fear.*
>
> *(Adorno & Horkheimer, 2002, p. 11)*

An Anti-Colonial (Sonic) Phenomenology of Spatialities Beyond Outside/Inside

Coloniality begins at a conjuncture of spatial disruption and is implemented with absolute (spatial) violence. As forceful implementation of *Naturbeherrschung*, the colonial/racial apparatus, in its various formations, performs (based on) a constriction of black life, violently reifying the Occident's curtailed dualistic conception of spatiality; resulting sites of wretchedness, in turn, provide the basis for falsely unambiguous abstractions.[16] Yet, the brutal mechanisms with which colonial power sustains itself collapse without the ontology of a divisible space. In this passage, we engage with the idea that there is a discrepancy between the prevalent Occidental conception of space as something divisible and how we experience it sonically, and how articulation of this discrepancy—as commonly enacted in black experiments with sound—is an anti-colonial option.

Raumvorstellung vs. Raumerfahrung[17]

Gaston Bachelard's *La Poétique de l'Espace* explores the confluence of spatial literary tropes, the production of space, and the constitution of experiential spaces. He outlines how modern dualistic philosophy constructs and sustains a strict division between outside and inside and produces, in this way, a certain conception of space (as something divisible). In his phenomenological study, Bachelard notes:

> Outside and inside form a dialectic of division, the obvious geometry of which blinds us as soon as we bring it into play in metaphorical domains. It has the sharpness of the dialectics of yes and no, which decides everything. Unless one is careful, it is made into a basis of images that govern all thoughts of positive and negative.
>
> *(Bachelard, 1994, p. 218)*

My interest in what Bachelard put forward lies in its poetic revealing of that which is beyond the seemingly natural, unchangeable intuition of a geometrically divisible space. Convincingly, the dialectic of outside and inside is, in his view, merely a pervasive metaphor, a "formal opposition" which construes space as something divisible, that has *defined* Occidental Thought since Euclid's system of geometric axioms, and in a more absolute way, since the Cartesian dualism—res cogito/res extensa. His rendering of (Bachelard, 1994, p. 212) geometric intuition as an artifice of the philosophical tradition makes it possible to assert other ways of thinking about spatiality and, consequently, as attempted here, options to undo racial/colonial space-taking. As the instructive basis for a re-description of spatiality beyond the traditional European dualism outside/inside, Bachelard cites the poetry of the nineteenth century; what we lean on here for this spatial unsettling is black sonic poetics/deformation/perforation.

The Coloniality of Geometric Spatiality

Under this caption one could provide a stringent, historically grounded analysis of how the simple geometric opposition outside/inside makes possible and is reinforced by the geospatial dispossessive methods and rationales of enclosure, property, occupation, territorial fragmentation and expansion, and racial-epidermal constriction that brought forth the modern world's colonial/racial formations; one could lay bare their continuities in contemporary economy, politics, and culture. Such an undertaking, however, would span several hundred pages and is not achievable here. Instead, the map (Figure 3.1) that shows contemporary national borders on the African continent, still haunted by those drawn during Bismarck's 1884/85 *Kongokonferenz* in Berlin, provides the backdrop for the following pages, which are intended to give a concise outline of how this geometrically restricted conception of spatiality is the ontological ground that underpins colonial/racial modernity.

FIGURE 3.1 *Map of Africa.*

Source: Illustration of the geometric division of Africa into fixed colonial states. Illustration. Freevectormaps. July, 17, 2022. https://freevectormaps.com/world-maps/africa/WRLD-AF-01-0002

The dialectic "outside/inside" promotes a conception of space as something to be taken, something to be occupied. Mentally picturing outside and inside as unambiguous dialectical opposites animates and is animated by a colonial unconscious. It is the ground (and the model) for the naturalisation of borders that allows a sharp partitioning of space into realms in which racialised (and gendered) claims to possession and sovereign power are made. With brute violence, racial/colonial modernity makes the metaphysics of outside/inside absolute.[18] Studies such as Brenna Bhandar's *Colonial Lives of Property*, (Bhandar, 2018) or Cherryl Harris' *Whiteness as Property* (Harris, 1993), show that irreducibly colonial logics of land ownership and anti-black racial regimes do not evolve on distinct, linear trajectories; they emerge from their interplay which—an idea I want to develop here —is corroborated by the unambiguous formal opposition of outside and inside.[19] Put differently, this spatial dualism is the immediate ontological ground for the demarcation of ownership of land, the expansion of territory, the nation-state, the fortification of borders, and also for the demarcation of who is deemed human and what is considered subhuman or a mere thing, culminating in the construction of sites of wretchedness and racial biopolitical (surveillance) technologies. The mechanisms of this multilateral process of (colonial/racial, epistemic, and geographic) escalation do not merely proceed from Europe's overdetermining eye; they also "feed" it.[20]

Les damnés as Ersatz Outside

If only briefly, I want to pick out one aspect of Sylvia Wynter's analysis of the epistemic rupture from the medieval to the modern epoch from her seminal "Unsettling the Coloniality of Being/Power/Truth/Freedom: Towards the Human, After Man, Its Overrepresentation—An Argument".[21] What we will concentrate on is how, despite the radical epistemic reorientation that is associated with the epochal break at the genesis of modernity, the basic idea of a dualistic spatial structure (inside/outside) is maintained and founds the figuration of the modern Subject vis-à-vis its irrational Other. It is worthwhile to outline Wynter's analysis because it allows us to (dis)place the spatial logic we have elaborated before (unambiguous distinction of outside/inside) as the ontological ground—or, perhaps, founding metaphor—of the modern world. As a central development in the construction of *Man* as the normative Subject, Wynter examines how, in the fifteenth and sixteenth centuries, renaissance humanists swept away two orthodox dualistic presuppositions, without which the medieval theocentric order and its derived conception of the human as a "True Christian Self" could not subsist: (Wynter, 2003). This was, first, the idea of a split geography of the earth, divided into divinely sanctioned, inhabitable regions (Europe) and divinely condemned, uninhabitable regions (Africa, "undiscovered" America), which was challenged by the European voyages to Africa and the Americas, disproving the supposed uninhabitability of those terrains (Wynter, 2003). The second orthodox presupposition the renaissance humanists—namely, Copernicus' heliocentric model—swept away was the idea of an ontological distinction between earth and the heavens.[22] After this spatial legitimising system of the medieval clergy had been intact for centuries, the Spanish/Portuguese voyages, along with the Copernican revolution, dismantled it by exposing the ontological distinction underpinning it as a fallacy, which opened up a *descriptive pause*, or, put in Wynter's terms, necessitated a new "descriptive statement". (Wynter, 2003, p. 264). The renaissance humanists, then, redescribed the human beyond the medieval theocentric conception of the human, now in terms of a distinction between the rational *Subject* and the irrational *Other*.

A crucial aspect of how Wynter reads this epistemic transformation is that, although the entire epistemic framework is restructured, the underpinning idea of a "space of Otherness" legitimises the paradigmatic conception of what it is to be human and is maintained and reaffirmed. (Wynter, 2003, p. 274). Wynter's reading is another approach to the absolute metaphysics of outside/inside that colours modern (colonial/racial) Thought: It is because the basic principle of "outside" as an ontologically distinct "space of Otherness" is upheld, even throughout radical epistemic ruptures in the renaissance, that redescribed "spaces of Otherness" are produced and projected onto those necessitated to be outside.[23] In other words, the underlying spatial dualism remains operative, even after the falsification of the medieval theocentric spatial ontological distinctions, and, thus, an *ersatz outside* is put in place. One could call this transformation a sublimation of the medieval ontological distinction, that is, from the theocentric spirit/flesh and heaven/earth to the post-medieval "degodded" (but not yet secular) ontological distinction—rational Subject/irrational Other (Wynter, 2003, p. 263). Wynter enables us to discern that the figure of *Man* as the normative Subject that emerges from the sustained and redescribed imaginary of an *outside* against which *Man* constitutes the new, only legitimate, mode of being human. "For the Europeans", says Wynter, "the only available slot of Otherness to their Norm, into which they could classify these non-European populations, was one that defined the latter in terms of their ostensible subhuman status (Wynter, 2003, p. 292). In other words, all that is *outside* of *Man* is turned into a "physical referent" (Wynter, 2003, p. 265) of *Man's* central position (Wynter, 2003, p. 265).

Construing outside/inside as opposites is an ontological artifice that dissects space into exclusive, non-interfering realms and, in this way, as Wynter's analysis shows, makes the "irrational savage" (dis)placed outside of history thinkable, giving rise to impassable borders as frontiers that mark malaise vs. prosperity.[24] Before we think about how sound offers a different approach to spatiality—unsettling racial/colonial intuition—I want to rewind to Bachelard who notes that

> … Metaphysics should … be resolutely discursive. It should beware of the privileges of evidence that are the property of geometrical intuition. Sight says too many things at one time. Being does not see itself. Perhaps it listens to itself.
>
> *(Bachelard, 1994, p. 215)*

Sonic Phenomenology of Spatiality

Sitting here as a black person socialised in Germany, it is not difficult to visualise an outside to the room I am in, the house I am in, the town, the country, the planet, and so on and so forth.[25] Such a *Vorstellung* of an outside is predominantly conditioned by the described visual, geometric approach to space.[26] Sound and hearing complicate this paradigmatic imagination. Although many have argued that sonic space is a less easily divisible space, with overlapping acoustic horizons and non-linear geometry, we do not need to rely on such treatises to get what is conducive to the argument made here.[27] Instead, it shall suffice to observe one's sonic surroundings. As I am writing this, for instance, I am sitting in a hotel room in Düsseldorf (ironically called *Hotel Bismarck)*; the windows and the door are closed, but still, I hear the residual hum of ventilator fans from other rooms, people moving down the hallway, birds chirping, or, occasionally, the siren of a police car racing by out on the main street. As sound, I take in spatial cues from all directions. Despite my being placed within an enclosed space, paying attention to what I can perceive aurally enables me to extrapolate the makeup of my environment, which emphasises my immersion into a space much larger than the confined structure I am in. Certainly, my mobility is immediately restricted by the walls of this room; however, I am also embedded in a perceptual transaction that exceeds its settled, non-acoustic spatial limit. In short, the conception (*Vorstellung*) of a segmented/divisible space is not coherent with my sonic, immersive experience of spatiality. There is something that exceeds simple geometric conceptions of space. In this way, I have become interested in the project of sonically unsettling the opposition outside/inside; put differently, my interest in the ontological/epistemic potential of black sound has become distinctly spatialised.

Articulating That Which Exceeds Modern Conceptions of Space

As we have seen, racial/colonial modernity reduces space to simple geometry, and this reduction inherently fails to consider that the sonic experience of spatiality offers a different ontology. One cannot begin to conceive of space, or land as something to be possessed, or even less, as a commodity, if we follow the sounds of black sociality to discard the ontology of a divisible space. To this end, we can think about, for example, the texture of trap, its heavy low-end, and how it unbounds Atlanta's clubs and, beyond any set horizons, permeates the entire globe. We can say, with the help of Simone White:

> The trap music producer exhales a pervasive—diffuse and dilute—sonic/affective atmosphere through the machines. Trap beat-making is a methodology of surround; so that we find ourselves in a club that we have not chosen to enter; though we have paid. The club is everywhere, and everyone is in it. It is put on the internet; it flies through the air.
>
> *(White, 2018, p. 142)*

In historico-cultural discourses, it is commonly accepted that aesthetic techniques such as perspective painting play a central role in solidifying the geometric conception of space as "natural", feeding modern intuition. Moreover, the invention of the printing press stiffened the West's insistence on the view that the production and transmission of stable, "scientific" and, as such, credible knowledge must wholly rely on visual representation. Heidegger infamously puts those developments at the threshold of the modern world in the following terms: "That the world becomes picture at all is what distinguishes the essence of modernity" (Heidegger, 2002, p. 68). The epistemic redescriptions that Wynter enabled us to discern are a result of this geometric picturing of spatiality. So, if innovations in technologies of seeing bring forth the paradigmatic ruptures from which modernity emerges, is it too far-fetched to conceive of the *black sonic* as a technology that disrupts the modern episteme, to think of the *perforating* 808 as our telescope, that heralds *post-dualistic* conceptions of space?[28]

Black sonic perforation opposes Dasein as schema. It is sounding that drowns out—in Wynter's terms—*Man's* dictate on "...how Self, Other and world should be represented and known". (Wynter, 2003, p. 268). On the following pages, we are going to look more specifically at what black sonic perforation implies, and how it unsettles sites of modern—which is to say racial/colonial—enclosure.

(Non)Performance, Perforation, and the Gesture of the Black Sonic (As Disruptive Noise)

Already on the first listen of keiyaA's *Forever, Ya Girl* I knew that this constellation of sounds would transform me. It was in September 2020. I remember vividly how I was completely immersed and that my fascination quickly carried over to the people I shared a home with. Four months later, I (virtually) attended a presentation keiyaA held at *Afrotectopia* entitled "Quickest Language: A brief into keiyaA's inspirations, creative approach and general philosophies around her recent work *Forever, Ya Girl*". Her talk focused on what she termed the tenets of black music. So, in between talking about polyphony, improvisation, repetition, and a communal listening of Stevie Wonder's "Flower Power", she uttered a phrase that has stuck with me, informing my thinking, writing, and sound-making since then. She said: "White supremacy thrives in silence".[29] With those words, she concisely encapsulated something that intuitively felt right (against the backdrop of the quiet countryside town I grew up in); it also felt something like an ethical imperative to disrupt this silence. It thrives if we do not call it out, but perhaps her phrase is also a warning that sometimes it can thrive *silently*.

What is central to our concern here is that black sonic interventions (as an ethical necessity, with epistemic implications) unsettle this silence, even unsettling the *architectures of constriction*, designed to guarantee and fortify this silence.

For a brief but crucial excursion to the contested soundscape of New York in 1919, we are following the unsurpassed "more-than-a-study" by Saidiya

Hartman: *Wayward Lives Beautiful Experiments*. In Hartman's own words, "[t]he wild idea that animated this book is that young black women were radical thinkers who tirelessly imagined other ways to live and never failed to consider how the world might be otherwise" (Hartman, 2019, p. xv). There is one specific scene of such radical imagination that we must dwell on. It is a riot enacted by young women imprisoned and tortured at Lowell Cottage, one of the cottages "reserved for coloured girls" at Bedford Hills Prison, New York (Hartman, 2019, p. 279) The following is an attempt to carefully attend to the sonic dimension of this riot which Hartman already articulates it in her chapter's title: "Riot and Refrain".

> The *New York Times* described the upheaval and resistance of Lowell Cottage as a sonic revolt, a "noise strike", the "din of an infernal chorus". Collectively the inmates had grown weary of gratuitous violence and being punished for trifles, so they sought retribution in noise and destruction. They tossed their mattresses, they broke windows, they set fires. Nearly everyone in the cottage was shouting and screaming and crying out to whomever would listen. They pounded the walls with their fists, finding a shared and steady rhythm that they hoped might topple the cottage, make the walls crumble, smash the cots, destroy the reformatory so that it would never be capable of holding another "innocent girl in the jailhouse". The "wailing, shrieking chorus" protested the conditions of the prison and insisted they had done nothing to justify confinement; they refused to be treated as if they were not human. The *New York Tribune* reported: "The noise was deafening…Almost every window of the cottage was crowded with Negro women who were shouting, angry and laughing hysterically". *"The uproarious din emanating from the cottage smote the ears of the investigators before they got within sight of the building"*. Songs and shouts were the instruments of struggle. Terms like "noise strike" and "vocal outbreak" described the soundscape of rebellion and refusal.
>
> *(Hartman, 2019, p. 279)*

This passage is difficult to excerpt, not only because it feels wrong to cut it from the singularly magnetic writing that precedes and succeeds it, but also because I have no way of soberly commenting on the situation we are placed in…

For sound to emanate from sites of radical enclosure, walls are deformed into ruinous markers of passage. The facticity of the walls faced by these young women, tortured and violated with brute physical force, is inevitable persons contained in a cramped cell, they are not supposed to be in. It is an enclosing falsehood, an enclosure of falsehood. Yet, in their raucous rendition of revolt, they deform the prison walls into an acoustic medium that will carry their resistive testament of the unjust site farther than its false façade.

With Hartman's words, we can say hauntingly, beautifully: "The chants and the cries escaped the confines of the prison even if their bodies did not" (Hartman, 2019, p. 284).

Plantocracy, carceral, or racial capitalism: Whatever you call this formation that is antiblack—that is, the only version there is—it ignites itself on a scheme of dismembering space, a scheme of constriction. As a disturbing (deforming) sonority, the raucous *black noise* that Hartman amplifies unsettles this plot. Its emanation is (our) evidence. "The chorus spoke with one voice"; this voice goes beyond outline (Hartman, 2019, p. 279).

Beyond outside and inside, the inside of such a space of captivity is a radical outside—or what May Ayim enables us to utter—an *us-enclosing outside*. May Ayim, "Das Jahr 1990: Heimat und Einheit aus afro-deutscher Perspektive", in *Entfernte Verbindungen: Rassismus, Antisemitismus, Klassenunterdrückung*, ed. Ika Hügel (Berlin: Orlanda Frauenverlag, Ayim, 1993), 208. At this point, we move from the sonic revolt against this state-sanctioned disciplinary space in New York to a sounding in May Ayim's domestic space in Münster, West Germany, in the late 1960s.

As I was writing the passage about "Riot and Refrain", I reengaged with May's autobiographical account in *Farbe bekennen* that carries the instructive title "*Aufbruch*".[30] In this intimate entry of little more than six pages, she lays bare her mostly troubled childhood growing up as the only black child under violent and restrictive (rarely loving) white foster parents (Ayim, 2020a, pp. 262). Their educational maxim was to forcefully ensure that May wouldn't match the negative cliches floating around about black Germans, which is to say, they were dismissive and constricted her.[31] As a nine-year-old, she desired to run away, to end her life. She wrote "*Aufbruch*" at 25, which, apart from describing her oppressive home, also documents her departure from it, carrying the affirmative voice of a young woman who has found her own path (Ayim, 2020a, p. 264). While I was rereading her account, I came across a passage that is worth placing alongside Hartman's writing on the riotous black noise pouring out from Lowell Cottage. Of her childhood's troubled years, May writes:

> Nicht aufmucken, lieber schlucken. Bis es nicht mehr geht und entweicht: ins Bett oder als brutaler Hustenkrampf, der jeden normal Hörenden zu schlaflosen Nächten und Wutanfällen treibt. So ist das mit der Unterdrückung.
>
> *(Ayim, 2020a, pp. 262–263)*

[Don't revolt, rather swallow. Until it breaks out: wetting the bed or as a brutal, convulsive cough, that drives anyone with normal hearing to sleepless nights and fury. Such is suppression/oppression].

To be constricted, cramped, yet "intoxicated with freedom", is it to live *bursting out*? (Hartman, 2019, p. 8). What stirs me here is how, underneath May's non-confrontative rationalisations *not* to revolt, it is as if her corporeal regulation initiates the revolt, and breaks the silence via an uproarious cough that projects her trauma, her (inner) out-of-placeness as a nighttime-disturbing *black noise*.

"These sounds travel through the night air", says Hartman about the riot at the torturous "reformatory". (Hartman, 2019, p. 286). Faced with the brutalisation of the matrons at Lowell Cottage, the young girls resisted with an intimate, communal, and belligerent performance of black noise. Faced with domestic violence and isolation, May's non-performance leads to an uncontrolled bursting out. It is not my intention to equate those two accounts or analyse in what ways they differ from each other. What is of interest here is how much the *perforative* moment of black sonic intervention is articulated in both accounts.[32] It is as if May Ayim's scene echoes the scene we encounter through Saidiya Hartman's writing, however, scaled down to the domestic, or even the somatic.

Black sonic perforation indicates the possibility of an intermingling of realms dissected by modern intuition (*Anschauung*) as ontologically distinct. Hartman documents this by citing the statement from the *New York Tribune*: "The uproarious din emanating from the cottage smote the ears of the investigators before they got within sight of the building" (Hartman, 2019, p. 279). Despite being in the most immediate form of confinement, the girls' sounding perforates the fortified walls, turning outside and inside into complexly blurry categories. Spatial dissonance is the result. Similarly, in May Ayim's account, the boundary that encloses her *cogito* (expressed as her rationale: "Don't revolt, rather swallow") is perforated and though she doesn't want to revolt, and her cough and her bedwetting are infuriating her dismissive foster parents even more, the disruption of their silence might be read as her very first steps towards an *Aufbruch*.

Both instances of noise-making emerge from a highly constricted and inflamed situation. We cannot insinuate that making noise is a linear option towards catharsis. Our focus should lie on the gesture of *perforation* that inheres in these black sonic interventions, fissuring markers of an outside/inside dualism. The moment of perforation we get to witness is animated by an (un)conscious rebellion, and projects *black noise* beyond confinement as a disturbance to *white silence*. It is this very gesture of perforation, the making possible of an *Aufbruch*, that is what we are attempting to study.

jazz ist ein schrei
der schweigen zerreiß.. (Ayim, 2013, p.188)

May Ayim, *jazz*

Miniatures from displace between

May Ayim
Her name is sound
solche schwingungen kündigen etwas an
2.275 black radiance
so she utters (something sublime)
May, here
as a, or the
historically influential, afro-german big sister
wholeheartedly voicing the silences History has left
silences that History keeps producing
silences in which their "supremacy" thrives
Her's is a fine voice
it is sometimes enclosed
hidden by this stifling
silence
it will not stifle you
yet again, yet another
echo of a voice that will
never decay
the unsettled place (dis place)
from which May speaks is one in which
fear of the outside is not worthwhile
in here
out here
physical contact with the outside is tolerable
even inevitable
inside is not kept secret;
the possibility to experience in/out as an ambiguous one is also
not kept secret
from a place
that touches and dissolves outside and inside
whose direction is in constant flux
she creates new possibilities:
Die Zwischenwelt als Chance—Toward a human after Man
our making of spaces
is not product
of an arbitrary phantasy
we are guided
sonorous voices
traces of existence

traces of the absent
Glissant let us know
History's claim to absoluteness is forever wrecked
and so we guard ourselves from the proper
and follow the traces they have left
. . . into the forest
and there we will rejoice and sing and say again:
the falling debris of the modern ruin is audible!

> Wir spüren, daß mit der bevorstehenden innerdeutschen Vereinigung eine zunehmende Abgrenzung nach außen einhergehen würde—ein Außen, das uns einschließen würde.
>
> —*May Ayim, Entfernte Verbindungen (Ayim, 1993, p. 208)*

> [We sense that the impending inner-German unification would be paralleled by an increased demarcation of an outside—an outside that would enclose us].

What fascinates me in this excerpt is the spatial dimension, and its underlying anti-geometric—or should I say *sonic*—principle that May registers. Like a wall-shattering sound, her disorienting figure of thought breaks from the geometrically opposed pair outside/inside. Pushing against this traditionally European xenophobic conceptualisation of space, we are reminded here of Adorno and Horkheimer's formulation that poses the Enlightenment as radical fear of the outside—May crafts the language for the peculiar and disturbing situation of being an Afro-German woman around the time of the fall of the Berlin Wall: A sustained and renewed relegation to the outside, a constricting exteriority.[33] The excerpt is from an essay entitled *Das Jahr 1990: Heimat und Einheit aus afro-deutscher Perspektive* ("1990: Home and Unity from an afro-German perspective") and anticipates how the addressed "increased demarcation of an outside", the renewed inner-German exclusion of "us" has, in the 32 years since then, too often erupted in violence.

Gegen xenophobes Raumdenken; Against xenophobic conceptions of space;

Black sociality as sonic intimacy?

(Glissant, 2010, p. 24)

> *Inner space is as infinitely explorable as spaces of the earth.*
>
> *(Glissant, 2010, p. 24)*

> *The eerie concerns the most fundamental metaphysical questions one could pose,*
>
> *questions to do with existence and non-existence: Why is there something here when there should be nothing? Why is there nothing here when there should be something?*
>
> *(Fisher, 2016, p. 12)*

—Mark Fisher, The Weird and the Eerie

"Absence"

When Glissant in his *Caribbean Discourse* famously speaks of a generative/poetic force that can piece together the fragments of a historical continuum shattered by the colonial order and its claims to universal History (Glissant, 1996, p. 62); it is my hope that this is equally true for the dis/continuum that exists as a function of the absence of black social life.[34]

In Riedlingen, where time seems to be passing slowly, where the Danube and green, forested hills are capable of providing a wholesome backdrop for a sheltered youth, my Dasein (*Abgeschottetsein)* is conditioned, besides abundant care and a pleasant, sometimes vivifying, stillness, by something more enigmatic.[35] It *is* that very stillness, and how it is too often condensed by the *weather of anti-blackness* to an austere, deafening silence(Sharpe, 2016, p. 106). Furthermore, my *Da-und-Dortsein* is conditioned by the palpability of a series of absences.[36] Only mediately—at least in my mind—I am talking about my father's absence; what I am immediately interested in is the absence of black social life.[37] Tina Campt points out "...what marks much of this [afro-german] community is the absence of the shared narratives of home, belonging, and community which sustain so many other Black communities and on which they draw...in numerous ways" (Campt, 2003, p. 290). What this passage allows us to articulate is that an Afro-German coming of age in a white, countryside town, can feel a bit like being stuck in a cascade of peripheries: A periphery to the white majority within the town, a periphery within Germany to a more cosmopolitan life in its cities; a periphery (lacking a present, unified community and culture) within the black diaspora. It occupies me how—at least in my experience—the permeation of black music makes a reorganisation of this spatial configuration possible.[38] Again, Simone White says:

> The trap music producer exhales a pervasive—diffuse and dilute—sonic/affective atmosphere through the machines. Trap beat-making is a methodology of surround; so that we find ourselves in a club that we have not chosen to enter, though we have paid. The club is everywhere, and everyone is in it. It is put on the internet; it flies through the air.
>
> *(White, 2018, p. 147)*

On my 15th birthday, my mother gifted me a silver MIDI controller. The production of Young Chop and Lex Luger mesmerised me and (re)awakened a desire to craft beats. In the early months of 2013, I discovered an interview with Young Chop in which he deconstructs his production for Chief Keef's *3Hunna* (Chop, 2012). Studying this video became a daily ritual. Soon, I started to engage and collaborate with other producers who were geographically remote, though virtually we congregated around. *flp* and *.wav* files; *those sounds (de)formed my (perception of) space, circulated my atmosphere.*

An Eerie Presence

Black music, as it permeates the rural geometrically divided landscape, enunciates something equally invigorating and upsetting: Invigorating because it evidences the seemingly limitless generative potential of black sonic idioms as a planetary force; upsetting because the presence of those sounds documented the radically alienating locus in which black sounds frequently exists as a mere entertaining background hum, as something for white Germans to "feel cool to", and consumed by them without any sense of self-reflexivity or acknowledgement of their positionality—that is (almost always) perpetuating anti-blackness.[39] As a permeating deformation, the black sonic makes the historicity of the space—its involvement in the project of racial capitalism and its haunting continuities—audible.[40] Christina Sharpe's ecological diagnosis that *anti-blackness is weather*[41] helps us to understand why—even in an eerie way, in a rural space, however spacious it may appear—the constriction that the racial/colonial order exerts over blackness is palpable.[42] To me, the environment feels dense because it forms a nearly impassable thicket of ignorance, xenophobia, *their* encroaching behaviour and—as curious as it might sound—me feeling at home/my feelings of home. Truly, it is a peculiar setting, and I am struggling to document it in this written form.[43]

In *The Weird and the Eerie*, Mark Fisher states: "The perspective of the eerie can give us access to the forces which govern mundane reality, but which are ordinarily obscured, just as it can give us access to spaces beyond mundane reality altogether" (Fisher, 2016, p. 13). Mechanisms of Modernity's constriction of black life (inseparable from xenophobic conceptualisations of space), that is, the unsayable eruptions of anti-black violence and targeted policing practices that structure (predominantly urban) life around the globe, eerily reverberate in a spacious environment like that of Riedlingen. Such reverberations are affirmed by its grid-like and standardised unnatural "natural" landscape and its stillness, yes, its beautiful, treacherous stillness.

A consequence of the perforation of unambiguous demarcations that we discussed earlier is that the rural and the urban cannot be conceived as isolated spheres: The pressure that is built up in the urban environment is not alleviated in the context of the spaciousness of the rural landscape. As weather, and here I evoke, again, Christina Sharpe, anti-blackness has long condensed into an ecological force, transgressing the *Man*-made borders between the rural and the urban... One could attempt to put it like this: To my experience of the rural in Germany, where black social life is "absent", the urban is a dialectical "outside"—where black sociality is possible, (and *premature death system immanent, always imminent*) (Sharpe, 2016, p. 132). Yet, as we know, black sound perforates distinctions between outside and inside, and in an eerie way, the idea of the "urban", its moving air, and its structuring of blackness, informs my experience of the rural.

> *The weird and the eerie [...] allow us to see the inside from the perspective of the outside.* —Mark Fisher, *The Weird and the Eerie.*
>
> *(Fisher, 2016, p. 10)*

Fields separated with utter accuracy, fences, *Man*-made lineaments, emblems fixed to trees indicating the bounds to different forest areas and properties are all experienced as indicators inseparable from the racial and colonial constriction of (terrestrial) life. In my immediate surroundings, this eerie geometry has sedimented. It is structuring the landscape—one of timber-framed houses, and perfectly mowed lawns organised into a grid. It only *appears* natural. The entire space, the entire landscape, condenses into a *Schauplatz* in which the material reality of racial capitalism plays out and becomes sensorially apprehensible.[44] When 808s thump and volleys of snare rolls hit in this landscape, it is a particular aesthetic experience, one of a radically constricting "outside". Here, black sociality is not in *sight*.[45] To be there, to exist here is to *trespass*.

Psychosomatic Drums

Ever since I can remember, I have been infatuated with the drum. In evocative ways, the drum can sonically carve out what *can't* be grasped about our surroundings—reaching farther than the *Begriff*.[46] Drum sounds were there, outside of the ones emanating from my mother's CDs—*Natty Dread*, *I'm Sorry Miss Jackson*, some Madonna tunes...—the ones I heard during the annual weeklong festivities before Lent—*Fastnacht*. After this short week, the rhythms I would hear from our local marching bands would dictate the rest of my year.

Before I could read or write, when my world was still mostly our house, I developed a habit—or better, the habit was just there, and it developed me. I would quickly wiggle my lower jaw left and right and grind my teeth to reproduce the rattling sound of the pacing snare drums they played in the marching band. Twitching my inner ear created a thumping with which I could counterpoint my teeth's snare with patterns of a pulsating bass drum.

Polyphonic drum patterns filled my head, resonated in my skull, and inner ear. Particularly in moments of nervousness, reverie, or concentration, I still frequently catch myself doing it.[47]

The Drum Skin and the Bread Knife

My sister gifted me a drum from the local "*Afrikaladen*"[48] called *ArtWareShop*, which was run, of course, by a guy called Jürgen. I don't remember much about this drum other than that it was a small version of a traditional West African drum, and that quite soon I wanted a different one, a "shiny" one, one you beat with sticks—like the one I had seen the boys play in the marching band. Finally, my mother gifted me a snare, ornamented with red and white triangles—perhaps, just after my fourth or fifth birthday—which gave way to a several-day-long flood of relentless drumming. On a Sunday shortly after, as always, my first steps led downstairs to pick up the snare drum. In an otherwise peaceful living room, illuminated by the pastel shimmer of the morning sun, the drum sat firmly on our upright piano, with its skin split open and a big, serrated bread knife pierced through. It was my mother's partner. Drumming without punctuation had enraged him, and he attempted to make an end to it while I was asleep. Of course, my mother, enraged by his act, got me the very same drum again.

No Trespassing

Alexander Weheliye, in his fundamental *Phonographies*, by way of Ralph Ellison's New York apartment, insightfully shows how, in urban zones, home speaker systems can produce enclosures of intimacy to shield off from city noise (Weheliye, 2005, p. 123). In the current undertaking, my curiosity lies in the aesthetics and the anti-colonial geospatial implications of the inverse movement: Sound that, in perforating walls, transgresses spaces of imposed enclosure. Perhaps this curiosity is attributable to my upbringing in a rural area where sound is more likely to trespass, rather than to be shielded.

Dualistic geometry is apparent not only in the rigorous partitioning of "property" but also in the attempted partitioning and disciplining of sonic space. Sound, however, always trespassing, diffuses over property lines. Throughout my youth, furious tenor solos, virtuosic sample collage wizardry, viscerally thumbing subs, or angelic vocal inflexions—whatever I was listening to, I intimately kept it contained within my headphones. Maybe this had its origin in a tender reminder that sometimes interrupted my drumming as a kid; my mother, approving of my noise-making, yet wanting to protect me from our neighbours' frowns, urged: "*Bis 14 Uhr ist Mittagsruhe*" [The midday quiet hours end at 2 p.m.] (Figure 3.2).[49]

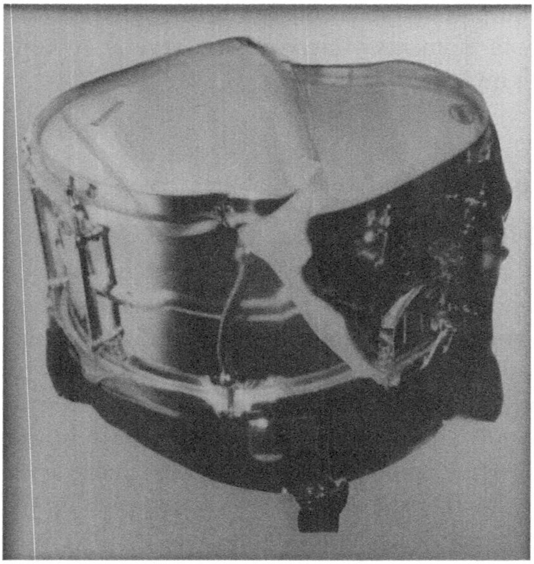

FIGURE 3.2 *Disfigured Polaroid of Drum*, Nathanael Kliebhan.

Source: From private collection. 2017.

Closing; Opening

Why write an essay about black sound without distinctly mentioning musical expression? Firstly, this essay was not an essay *on* black sound, as much as it was an essay on the unsettling of an unambiguous opposition of outside and inside, thinking *with* and *through* black sound, in which this performative *gesture*, faced with racial/colonial constriction, is commonly practised. Certainly, as a result of centring black sonic perforation it would have been appropriate to talk about the sublime bass frequencies of dub and sound-system culture, or, to talk about black electronic music more directly, however, my curiosity was directed towards using textual excerpts that—equally sonorous and perforating—shatter walls in their own (more or less) intimate ways, while being drowned out by discourses on music. This was also to attempt to follow Saidiya Hartman's centring of "young black women [who] were radical thinkers who tirelessly imagined other ways to live and never failed to consider how the world might be otherwise" (Hartman, 2019, p. xv).

Besides, black music *has* been an integral part of my methodology and writing. When, after an extended writing session, my head was buzzing, I took breaks in which I put on music (see appendix), cranking up the volume as loud as possible, or tried to figure out some chords on the acoustic guitar. I am writing as a musician and sound-maker, and the composition of this text has been in many ways so identical to heartfelt musical composition that I am struggling to tell those two supposedly different modes of output apart. Therefore, after these pages dedicated to the unsettling of borders, a final perforation might be that of the distinction between theory and practice. This undertaking was but a single step towards embodying this integration.

To conclude, after a problematisation of the impossibility to define black sound, we have engaged with *deformation* and *perforation*, to describe the black sonic as a permeating oscillation with a critical, that is, anti-colonial, anti-anti-black impetus. Subsequently, we looked at *Naturbeherrschung* as a radically reductive project of enclosure and division that *includes* the domination of the human. Theo Parrish's astute point that techno uses the machine as an escape from the process in which the assembly line assimilates the human demonstrated that a black sonic deformation of technology (techno) *is* against *Naturbeherrschung*. We engaged with Bachelard's proposal that there is an experience of spatiality which is beyond the intuition of a dualism of outside/inside, rendering the idea of a geometrically divisible space a curtailed ontology. From this, we outlined the outside/inside dialectic as the ontological ground of racial/colonial modernity, alongside which Wynter's analysis of the epistemic ruptures at the genesis of Modernity, showed that the figure of the modern Subject emerges against the ground of a constant "space of Otherness"; substitute for the medieval ontological distinctions was the newly inscribed irrational (Native/African) Other. Unsettling this geometry, a reading of Saidiya Hartman's incomparable "Riot and Refrain" allowed us to trace the performative chorus that the girls imprisoned at Lowell Cottage exhaled to transcend their confinement. Such sonorous breaking out is paralleled in May Ayim's "Aufbruch", which completed the sonic unsettling of the outside/inside dualism. In "Miniatures from displace between", I put forward a short poem and engaged with a citation from Ayim's essay on her disorienting experience after the fall of the Berlin Wall. With the help of Christina Sharpe's formulation "anti-blackness as weather" and with a loose integration of fragments from Mark Fisher's *The Weird and The Eerie, the* last passage centred my experience in Riedlingen in relation to its geometric space, and questions on the "absence" of black social life and its "re-presentation" through black musical permeation.

Ideally, the preceding pages would be an articulation made possible by *black sonic Bildung*.[50] To fully unpack what this process means, another extensive study is necessary, or perhaps it suffices to carefully listen to the *music that is floating, out there, in here*. Still, I think it is fruitful to give some hints as an opening. Dualistic Thought and the kind of spatiality it suggests emerge from the modern symbolic order, for which there is no space for ambiguity; it strives towards disambiguation. Yet, as the phenomenology of sonic space and the gesture of black sonic perforation have shown, how space is to be inhabited cannot be exclusively pictured, nor can it be derived from its geometric representations. Coloniality (personified as the *settled* Self that extracts itself from the natural continuum and locates itself in a Cartesian continuum) stubbornly materialises and institutionalises a constricting geometry whose demarcations cross biotopes and Thought. In a lecture entitled "Building and Bildung und Blackness: Some Architectural Questions for Fela" Fred Moten states:

> The ability…to stand one's ground has been so crucial to this notion of picturing … One has a picture of oneself, one has a sense of oneself, precisely to the extent that one can be locatable. …This is how the construction of the modern, normative Self is so bound up with what it is to own, to occupy land, to settle, to take up space.
>
> *(Moten, 2022)*

Such already colonial, already racial *Bildung* of Self bound up in "what it is to own, to occupy land, to settle, to take up space" is unbounded and unsettled by the sounds we have described. Opposed to *Bildung* as *picturing*, they are expressions of alternative, sonic modes of *Bildung*—one could say Self-*less*-formation. With *black sonic Bildung*, a post-dualistic formation, that is, a *building* and being where bounds between inside and outside are perforated, one learns to bear/allow/practice/foster/cultivate a polyphonic dialogue within, as an echo of our experiential world(s). Black sonic Bildung is beyond the divisive gathering of geometric coordinates of a visually represented world, a *sonic* gathering of experience, and a sonorous *gathering* of *us*. Here and there we are. Animated by the world of black sound, this *Bildung* is building—*we* are building—something less violently cramping and something more inhabitable than the modern formations of enclosure. Black sonic Bildung, this deformed formation and building, and deforming being, is *Da- und Dortsein* despite the imposed enclosure of Europe's gaze; it is, quite openly, to be at home in the obscurity of sound.

Notes

1 Sometimes, as in Adorno's and Horkheimer's case, what they *fail* to think is source of such fascination.
2 Expression of this unsettled position might be that, more than I usually aim for, throughout this text some words—if they enunciate something particular—will be employed in German. In all cases, I have provided a translation.
3 Both the language used in this writing, as well as the gesture at its centre (the deforming black sound that perforates walls and brings forth fruitful complexities), are *not* to be thought of merely as metaphoric play. We know the crisis is real, and its violence absolute (the beauty is too). What will be said and how I attempt to formulate it emerges from an urgency, which, in-turn, results from the continual experience of a pull—or, to speak with Christina Sharpe, from the

experience of being *in the wake*. See, Christina Sharpe, *In the Wake: On Blackness and Being* (Durham: Duke University Press, 2016), 3.

4 Alexander Weheliye articulates this in his exceptional reading of Ralph Ellison's "Living with Music":

> Ellison...reveals that sonic technologies provide one of the central ways for subjects to negotiate shifting sonic public/private perimeters in cities, showing that proliferating noises can be kept at bay by establishing one's own sonic perimeters, all the while contributing to the burgeoning of these sounds.

See, Alexander Weheliye, *Phonographies: Grooves in Sonic Afro-Modernity* (Durham: Duke University Press, 2005), 143.

5 The offerings of Georgia Anne Muldrow, Lee Perry, Madlib, Moor Mother, Lex Luger, Matana Roberts, Andre 3000, Nina Simone, Elvin Jones, Missy Elliott, and many others enable this writing; however, I do not want to write *about* the content of acoustic worlds they craft, simply because my writing is inadequate in relation to those worlds. I do not want to hermeneutically read their virtuosic interventions and recordings as cultural texts. Instead, my interest lies in the kinetic/philosophical modalities of sound, the material effect of *kinopoiesis*, and how it effects being in/with this world.

6 Wynter's instructive reading and brilliant extension of Foucault's concept of the "invention of man" will support this further down.

7 *Man* is used throughout this account *not* as a cypher for humankind, but in the Wynterian sense, that is, as a constructed Eurocentric heteropatriarchal figure that emerges from the epistemic shifts at the origin of the modern world. This is marked by capitalisation and italics. A brief engagement with Wynter's analysis will hopefully make this more apparent.

8 We will engage with the metaphysics of colonial spatial concepts more thoroughly; this short paragraph is just to immediately link it to the domination of nature.

9 It was not intended to distinctly write about the Subject; however, during the course of this research, the Subject has increasingly presented itself to be an untenable figure. Throughout this text, it is only possible to give a cursory glance at black sound's *dis figuration*.

10 It might be necessary to note that there is a dissonance involving Adorno, with his polemics on jazz and his incapacity to discern the critical potential of black aesthetic practice. Nonetheless in a black deformation of his astute, yet, entirely Eurocentric critical output I see a fruitful complexity. For detailed critical study see, Fumi Okiji, *Jazz as Critique: Adorno and Black Expression Revisited* (Stanford: Stanford University Press, Okiji, 2018).

11 "What human beings seek to learn from nature is how to use it to dominate wholly both it and human beings. Nothing else counts". Max Horkheimer, and Theodor W. Adorno, *Dialectic of Enlightenment: Philosophical Fragments* (Stanford: Stanford University Press, [1947] Horkheimer & Adorno, 2002), 2.

12 They state: "The myth desired to report, name, to find the origin, hence, also to depict, record, and explain". See,[20] Horkheimer, Adorno, *Dialektik der Aufklärung*, 14 [my translation].

13 "Corrosive rationality" is my translation of *zersetzende Rationalität*. See, Horkheimer, Adorno, *Dialektik der* [21] *Aufklärung*, 12.

14 In "Blackness and Poetry", Fred Moten notes: "... modernity (the confluence of the slave trade, settler colonialism and the democratization of sovereignty through which the world is imaged, graphed and grasped) is a socioecological disaster ... ". See, Fred Moten, "Blackness and Poetry", *Evening Will Come: A Monthly Journal of Poetics*, issue 55 (July 2014) (Moten, 2014) [text without page numbers].

15 In *Wretched of the Earth*, Frantz Fanon notes "[…] Marxist analysis should always be slightly stretched every time we have to do with the colonial problem". See, *Wretched of the Earth* (New York: Grove Press, [1961] Fanon, 1963), 40.
16 Unfortunately, there are too many examples: the barracoon, hold, plantation, colony, slum, ghetto, prison…
17 This translates to "the visual representation of space vs. the experience of space".
18 I have already written about this concern in my previous account, *Initial Observations on The Dark Continuum.* The logic of the divisibility of space manifests in the colonial idea of the establishment (and expansion) of territory and political sovereignty. In this previous account, I referenced the spatial analysis of Achille Mbembe:

> Colonial occupation itself consisted in seizing, delimiting, and asserting control over a geographical area—of writing a new set of social and spatial relations on the ground. The writing of new spatial relations (territorialization) ultimately amounted to the production of boundaries and hierarchies, zones and enclaves; the subversion of existing property arrangements; the differential classification of people; resource extraction; and, finally, the manufacturing of a large reservoir of cultural imaginaries.

See, Achille Mbembe, *Necropolitics* (Durham: Duke University Press, 2019), 79.
19 See, Brenna Bhandar, *Colonial Lives of Property: Law, Land, and Racial Regimes of Ownership* (Durham: Duke University Press, 2018) and Cheryl Harris, "Whiteness as Property", *Harvard Law Review* 8, vol. 106 (June, 1993): 1707–1791.
20 One can, perhaps, best corroborate this with the following quote from Frantz Fanon: "And thus it happened that we encountered the white gaze. An unusual heaviness constricted us". See, Frantz Fanon, *Schwarze Haut, weiße* Masken (Berlin: Turia + Kant, [1952] Fanon, 2013), 94 [my translation].
21 Note that due to the constraints of this essay it is not possible to fully trace the framework of Sylvia Wynter's vast engagements in "Unsettling the Coloniality of Being". Even less than a severely abridged outline, this is an attempt to highlight a single point; namely, the outside/inside dialectic is a continuous factor underpinning the epistemic breaks Wynter analyses.
22 Copernicus' worldview was an "…intellectual challenge to the ontological distinction between the supralunar and sublunar realms of the cosmos [and] to its foundational premise of a nonhomogeneity of substance between them". See, Wynter, "Unsettling the Coloniality of Being", 277.
23 "Necessitated" is meant in the sense that those "spaces of Otherness" are needed to stabilise the episteme, the socio-economic/material conditions and to give a (negative) answer to the question of what it means to be human.
24 Forgive me for not being able to omit this cursory remark: The Fanonian *racial epidermis* must be viewed as one of such frontiers. See, Frantz Fanon, *Black Skin, White Masks* (New York: Grove Press, [1952], Fanon, 2008), 92f.
25 I chose to write this strictly from my phenomenological perspective, to stay away from formulations that have a universalist (ableist) ring.
26 *Vorstellung* is an imaginary, image, that is *set before one's eye.*
27 See, for instance, R. Murray Schafer, *The Soundscape: Our Sonic Environment and the Tuning of the World* (Rochester: Destiny Books, [1977], Schafer, 1994), 81.
28 The 808 is used here merely as a cypher; that such an unsettling of Occidental spatial dualisms can emerge from various kinds of *black sonic perforation* will become evident in the next passage.
29 Unfortunately, this session was not recorded.
30 The noun *Aufbruch* translates as "fissure", "tear", but also as "departure", "awakening".

31 May reflects: "Angst gab es genug. Wahrscheinlich Platzangst. Oder Angst zu platzen. [There was plenty of fear. Perhaps it was fear of constriction or fear of bursting from it.]" And identifies:

> I grew up with the sentiment, that was actually theirs: having to prove that a "half breed", a "negro", a "child from the protectory" is a full human being. Hence, there was barely time and space to discover who "I" was.

See, Ayim, "Aufbruch", 262, 266 [my translation].

32 This is not to be read as me turning May's troubled childhood into a spectacle, or construing the scene she describes as an overt site of resistance; equally, I do not want to compare the torturous experience of the girls at Lowell Cottage to any kind of domestic situation. Hartman is clear: "No one in Lowell Cottage mistook it for a home." See, Hartman, 268. Evading these clumsy suggestions, I want us to carefully hold the idea that there *is* a parallel between the situation that Hartman and Ayim articulate, namely, their *perforative* gesture.

33 In this account, I am using "Afro-German" and "black German" interchangeably. This is signified by the lower case "Afro-German" instead of the more common "Afro-German".

34 There are open questions about this absence: Where/when/how is it possible to (re)present black sociality through black sound? Why/how/in what contexts does one aestheticise the absence of black sociality? Is black sociality ever *wholly* absent? I am listening.

35 One could translate this as "Being-isolated".

36 This formulation which could be translated as "Being-here-and-there" might be a more appropriate ontological descriptor than mere *Dasein*.

37 One could say that those absences are the same thing; is it pars pro toto or toto pro pars?

38 The absence of black social life and its "re-presentation through" black sound—remains a perpetual hypothesis.

39 While writing this my inner ear kept circulating this quote by scholar and composer George Lewis: "Listening turns out to be different from just hearing. The only way you really know if you're being listened to is by the response". See, "George Lewis on Composing Behaviors", *YouTube video*, 00:01, by *Doris Duke Charitable Foundation*, April, 2020, Lewis, 2020, https://www.youtube.com/watch?v=zvzq2ffdMMM&t=57s

40 Again, where/when/how is it possible to (re)present black sociality through black sound? Why/how/in what contexts does one aestheticise the absence of black sociality? Is black sociality ever *wholly* absent? I am listening.

41 Sharpe states: "In what I am calling the weather, antiblackness is pervasive *as* climate. The weather necessitates changeability and improvisation; it is the atmospheric condition of time and place; it produces new ecologies". See, Sharpe, *In the Wake*, 106.

42 Of course, this is intermingled with the haunting presence of Germany's Nazism.

43 In an ongoing series of compositions entitled *trespassing*, I am attempting to document this sonically.

44 *Schauplatz* can be translated as "site of gazing". It has a different ring than the English "showplace".

45 And again—where/when/how is it possible to (re)present black sociality through black sound? Why/how/in what contexts does one aestheticise the absence of black sociality? Is black sociality ever *wholly* absent? I am listening.

46 *Begriff* translates to "term". In a literal sense, it invokes "grip" or "grasp".

47 How do we bear/allow/practice/foster/cultivate a polyphonic dialogue within, as an echo of our experiential world(s)?

48 This is a shop that sells "African" goods.
49 [The midday quiet hours end at 2 pm.]
50 Bildung describes the process of formation/education. It evokes the Bild (picture), but also the Gebilde (building, formation, and landscape).

References

Adorno, T.W. and Horkheimer, M. (2002 [1944]) *Dialektik der Aufklärung: Philosophische Fragmente*. Frankfurt am Main: Fischer Verlag.

Ayim, May. "Das Jahr 1990: Heimat und Einheit aus afro-deutscher Perspektive." In *Entfernte Verbindungen: Rassismus, Antisemitismus, Klassenunterdrückung*. Edited by Ika Hügel, 206–222. Berlin: Orlanda Frauenverlag, 1993.

Ayim, May. *Weitergehen: Gedichte*. Berlin: Orlanda Frauenverlag, (1997) 2013.

Ayim, May. "Aufbruch." In *Farbe bekennen: Afro-deutsche Frauen auf den Spuren ihrer Geschichte*. Edited by M. Ayim, K. Oguntoye, and D. Schultz, 261–267. Berlin: Orlanda Frauenverlag, (1986) 2020a.

Ayim, May. "Rassismus hier und heute." In *Farbe bekennen: Afro-deutsche Frauen auf den Spuren ihrer Geschichte*. Edited by M. Ayim, K. Oguntoye, and D. Schultz, 190–196. Berlin: Orlanda Frauenverlag, (1986) 2020b.

Bachelard, Gaston. *Poetics of Space*. Boston: Beacon Press, (1958) 1994.

Benjamin, Walter. "Über einige Motive bei Baudelaire." In *Walter Benjamin Gesammelte Schriften I*. Edited by Rolf Tiedemann, and Hermann Schweppenhäuser, 607–653. Frankfurt am Main: Suhrkamp, (1939) 1991.

Bhandar, Brenna. *Colonial Lives of Property: Law, Land, and Racial Regimes of Ownership*. Durham: Duke University Press, 2018.

Campt, Tina. "Reading the Black German Experience—An Introduction." *Callaloo* 26, no. 2 (Spring 2003): 288–294.

Chop, Young. "Off the Record: Young Chop." *YouTube video*, by *LRG Clothing*, April, 2012. https://www.youtube.com/watch?v=whvS70vlARM.

Fanon, Frantz. *Wretched of the Earth*. New York: Grove Press, (1961) 1963.

Fanon, Frantz. *Black Skin, White Masks*. New York: Grove Press, (1952) 2008.

Fanon, Frantz. *Schwarze Haut, weiße Masken*. Berlin: Turia + Kant, (1952) 2013.

Fisher, Mark. *The Weird and the Eerie*. London: Repeater, 2016.

Glissant, Édouard. *Caribbean Discourse*. Charlottesville: University of Virginia Press, (1981) 1996.

Glissant, Édouard. *Poetics of Relation*. Ann Arbour: University of Michigan Press, (1990) 2010.

Glissant, Édouard. *Introduction to a Poetics of Diversity*. Liverpool: Liverpool University Press, (1996) 2020.

Gumbs, Alexis Pauline. *Undrowned: Black Feminist Lessons from Marine Mammals*. Chico: AK Press, 2020.

Harney, Stefano and Moten, Fred. *The Undercommons: Fugitive Planning & Black Study*. New York: Minor Compositions, 2013.

Harney, Stefano and Moten, Fred. *All Incomplete*. New York: Minor Compositions, 2021. 42.

Harris, Cheryl. "Whiteness as Property." *Harvard Law Review 8*, 106 (June, 1993): 1707–1791.

Hartman, Saidiya. *Wayward Lives, Beautiful Experiments: Intimate Histories of Social Upheaval*. London: Serpent's Tail, 2019.

Heidegger, Martin. *Off the Beaten Track*. Cambridge: Cambridge University Press, (1950) 2002.

Horkheimer, Max and Adorno, Theodor W. *Dialektik der Aufklärung: Philosophische Fragmente*. Frankfurt am Main: Fischer, (1947) 1988.

Horkheimer, Max and Adorno, Theodor W. *Dialectic of Enlightenment: Philosophical Fragments*. Stanford: Stanford University Press, (1947) 2002.
Lewis, George. "George Lewis on Composing Behaviors." *YouTube video*, 00:01, by *Doris Duke Charitable Foundation*, April, 2020. https://www.youtube.com/watch? v=zvzq2ffdMMM&t=57s.
Moten, F. (2018) *The Universal Machine: consent not to be a single being*. Durham, NC and London: Duke University Press.
Moten, Fred. "Blackness and Poetry." *Evening Will Come: A Monthly Journal of Poetics*, no 55 (July 2014/2015).
Moten, Fred. "Building and Bildung und Blackness: Some Architectural Questions for Fela." *YouTube Video*, 34:10, by *MIT Architecture*, April, 2022. https://www.youtube.com/ watch?v=EW1aR3rHeZE&t=3609s
Okiji, Fumi. *Jazz as Critique: Adorno and Black Expression Revisited*. Stanford: Stanford University Press, 2018.
Parrish, Theo. "Feature: Theo Parrish." *YouTube video*, 9:24, by *Telekom Electronic Beats*, May, 2012. https://www.youtube.com/watch?v=-5USZQ97l9s&t=564s
Schafer, R. Murray. *The Soundscape: Our Sonic Environment and the Tuning of the World*. Rochester: Destiny Books, (1977) 1994.
Sharpe, Christina. *In the Wake: On Blackness and Being*. Durham: Duke University Press, 2016.
The Roots. "Quicksand Millennium." *21 Things Fall Apart (Deluxe Edition)*. MCA, 1999, CD.
Weheliye, Alexander G. *Phonographies: Grooves in Sonic Afro-Modernity*. Durham: Duke University Press, 2005.
White, Simone. *Dear Angel of Death*. Brooklyn, NY: Ugly Duckling Presse, 2018.
Wynter, Sylvia. "Unsettling the Coloniality of Being/Power/Truth/Freedom: Towards the Human, After Man, Its Overrepresentation—An Argument." *CR: The New Centennial Review* 3, no. 3, (Fall 2003): 257–337.

Appendix

John Coltrane—Your Lady
Ambrose Akinmusire—Umteyo (Live), Taymoor's World (Live)
RP Boo—Area 72, Speakers R-4
Jana Rush—Fluidity (Cerebral)
Loraine James—Loraine James
Lee Perry—Jah, Crab Yars
Robert Hood—Home
Wacka Flocka Flame—Damn
Moor Mother—Black Flight (feat. Saul Williams)
Patrice Rushen—Where There Is Love
Latrelle—Infatuated
Niecy Blues—Keep Goin', Black Girl Ur Not Far, Bones Become the Trees
keiyaA—Nu World Burdens, Rectifiya
Robert Glasper—Mood
The Internet—Red Ballon
Thundercat—Seasons

4

RE-SITUATING MACHINE LISTENING

The Question of "Justice" in the Voice Biometry Apparatus of the German Asylum System

Pedro JS Vieira de Oliveira and Shintaro Miyazaki

Introduction

In this chapter, we offer a proposition for critically assessing and re-situating the term "machine listening" and its applications in voice biometry. Following current thinking in sound studies scholarship that criticises the modern/colonial framework of listening (Thompson 2017; Goh 2017; Robinson 2020; de Oliveira and Pedro 2020), we outline how machine listening exacerbates and appropriates modern/colonial assumptions about listening and, more specifically, what this listening is claimed to produce. While it is beyond the scope of this chapter to provide a historiography of the term "machine listening", it is important to emphasise that the processes found in early histories of speaker identification and recognition inform more recent, automated methods of voice biometry.

Furthermore, we take issue with the idea that "machine listening" is fundamentally different from human listening – and in so being, more "neutral" or "objective" – offering instead a reading of machine listening and human listening as essentially co-constitutive activities. Consequently, we argue that there is an inherent contradiction in using "listening" as both a placeholder and a metaphor for feature extraction: while it implies, on the one hand, that machines are suited to performing tasks that depend on subjectivity (interpretation, affect), at the same time, it sediments the notion that listening (human or otherwise) can produce a "truth" about a body. We argue, instead, that there is as much "humanness" in machine listening as there is in "human" listening itself. This sort of humanness, although designed to render itself and its subject invisible, is fundamental to understanding that machine listening is, in its genesis and attribution, a colonial and colonising device. This is

DOI: 10.4324/9781003194453-5

precisely because the nature of machine listening is ontologically similar to "human listening" practices.

At its most simplistic level, biometry can be defined as pertaining to the measurement of the body; yet to define exactly where a "measurement of the body" begins or ends requires more than an etymological parsing of the word and must, necessarily, include a historical analysis of the desire to render the body measurable. Tracing the historicity of measuring otherness establishes biometric technology as intrinsically connected with the desire not only to read difference but also to define the ways in which that difference endorses categorical and political exclusion. We thus define biometric technologies as encompassing not only how the body is measured and evaluated but also what these measurements do to a body, what kind of body they produce, as well as where this production becomes legible; that is, biometry "is constituted by the practices involved in its use" (Murray 2007, p. 349). While most scholarship on biometry concerns visual markers (fingerprints, signatures, irises, scars, birthmarks, and so on), our intention in this chapter is to extend the problematisation of biometry to its aural and acoustic dimension, in particular to the use of voice as a biometric marker. Although other sonic markers do exist, voice seems to be the main locus of biometric extraction.[1]

Acts of collecting, taxonomising, and deriving measurements from sounds produced by the body also originate from anthropometry and eugenics, both tenets and principles of the so-called "racial sciences", and thus partake not only in the violent history of colonisation and capture, or in making the racialised body legible and visible to the state (Browne 2015, pp. 65–74), but also in its continuation in border regimes (see, e.g., Amoore 2006; Keshavarz 2019). In that regard, we raise specific concerns with (a) how biometric technologies are designed with the assumption that voices are not only physically but also discursively anchored in the body, (b) the use of "machine listening" as the main operativity with which to render that assumption traceable and therefore believable, and (c) how "voice" is further weaponised in border regimes and practices with the purpose of regulating national identity, citizenship, and the right to belong. The articulation of these three concerns is what we regard as constitutive of "voice biometry".

Critical scholarship on biometry quite correctly affirms that, among its many problematic stances, an underlying theme is its attempt to fix and secure identity "as a basis for prediction and prevention" (Amoore 2006, p. 339). Biometric technologies are designed and tailored to eliminate (or obscure) fuzzy boundaries to increase "objectivity", tying identity to the body. However, as we will argue throughout this chapter, this fixity is not a product of biometry but instead a continuation of an instrumentalised, utilitarian, colonial, racialised orientation towards identification as a form of resolution and truth-finding. Thus, we first highlight the necessity to critically assess listening and decouple voice, body, and identity as constitutive of one another. We focus our inquiry on a set of practices in which these orientations might be not only more identifiable but also convey the strongest consequences: the application of machine listening and voice biometry in the context of migration, border enforcement, and asylum law. We maintain that precisely because of the lexicon upon which machine listening is defined and constructed, that is, listening as categorisation, separation, and resolution, voice biometry is implicated in a chain of representations ascribed with increasing ontological power. It participates in a system invested in reifying rationalist, utilitarian, and even racist ideas that measurements can stand for voice, and thereby can stand for someone's identity.

Producing Aural Difference

Listening was one aspect of a set of epistemological apparatuses employed in the "civilising" process of colonisation. Early colonial practices of scientific listening and documentation are situated precisely at the boundaries between what Europeans perceived to be "human" and their "uncontainable", less- or non-human counterparts. As a documentation practice or disciplinary device, listening to the sounds of Europe's "others" stripped non-Europeans of their own "proprioceptive agency" (Robinson 2020, p. 55), not only in their own lands but also within Europe's wars and the borders they created. Written descriptions of listening, and, later on, recordings produced mostly by Northern European colonial travellers and "explorers" in the Americas significantly helped shape the so-called "natural sciences" and the institutionalisation of scientific methods and were thus "crucial to the construction of the epistemic transformations of modernity" (Ochoa and Maria 2014, pp. 36–37). Similarly, the indigenous peoples of the Americas were thought to lack the capacity of "proper" listening and were thus subjected to corrective regimes of pitch, harmony, and tempo (Robinson 2020, p. 54) while their own voices and sovereign practices were extracted and recorded for preservation in the face of "a false ascription of the inevitable extinction of Indigenous cultural practices" (Robinson 2020, p. 153). Therefore, modern/colonial forms of listening sediment a desire to reduce contingency, unravelling an ontology of extractive listening oriented towards rationalisation and categorisation, progression, resolution, familiarity, certainty, and fixity (Robinson 2020, 50–51) – both as correction and as congealment.[2]

The recorded voice, accordingly, appeared in the late nineteenth-century as another source of "scientific listening" stemming from colonial observation and documentation, this time fusing state-of-the-art technology with scientific racism, anthropometry, and physiognomy to mark, categorise, and differentiate individuals and groups of individuals. Early histories of sound recording from Europe and the US suggest a specific concern with the voice as a source of evidence and as a method for identification and classification.

Recorded voices were attributed a revealing power, able to expose characteristics of an individual that would otherwise be obscured by the recording medium itself – such as age, origin, gender, and even hair colour (Li and Mills 2019). Nonetheless, they were also thought to betray the speaking body by revealing hesitation, falsehood, and deceit. The belief in the ability of sound (and the voice) to reveal "truth" about the speaker was not limited to identification techniques of listening; it also required the development of specific listening practices to be acquired and performed initially by skilled listeners, and, later, by machines. The phonograph enabled novel techniques for the assessment of sound and voice by means of recording and reproduction, extending the ability of the skilled listener to perform multiple hearings of the same voice and, in turn, to produce an informed, scientifically accurate evaluation of its speaker.

A good example is found in early-twentieth-century Northern Europe, where phonetician Wilhelm Dögen, who led the Royal Prussian Phonographic Commission, travelled across most of today's Germany with a phonograph and a team of 50 employees to produce recordings of prisoners of war. His intended goal was to assemble a comprehensive library of accents and dialects with the aim of analysing and understanding the "cultures" of the world (Dögen 1925, p. 9).[3] The Commission's recordings were accompanied by annotated documents (*Personalbogen*) outlining national, linguistic, and familial relations, as well as articulatory diagrams and phonetic transcriptions of the recording. A final evaluation (*Urteil*) of the recorded person was then produced by the linguist in charge – often Dögen himself (Lautarchiv n.d.). These recordings not only helped constitute the *Lautabteilung* of the Prussian Library in Berlin but also yielded an extensive visual and written documentation titled *Unter fremden Völkern* ("Among Foreign Peoples"), published in 1925. In the following years, Dögen would extend his interest to recordings of convicted men, working on the assumption that there could be a fundamental connection between voice and criminality (Li and Mills 2019, p. 130).

Indeed, for the German linguists, phoneticians, and naturalists of the nineteenth and early twentieth century, observation not only aided but also directly informed listening, whereby "sonic perception [was] spread on corporeal difference, scientific explanation, and the narration of uncontainable, bodily produced noises" (Ochoa and Maria 2014, p. 32). Dögen's interest in producing visual material from those individuals he forcefully recorded in prisoner of war camps was connected to the possibility of its correlation with the linguistic articulation of dialects and accents, and in particular with the "rare noises" made by "exotic" groups (Dögen 1925, p. 16). That many of these prisoners of war were also colonial subjects, stationed in Europe fighting their coloniser's wars, seemed to offer an unprecedented opportunity not only to record the languages of non-Europeans but also to validate Germany's belief in racial hierarchies (Dögen 1925, p. 15). Unsurprisingly, the wording employed by Dögen to describe the ethnic groups he and his commission recorded relies heavily on eugenicist and racist terminologies and need not be reproduced in this chapter. What is interesting to note, however, is their dependence on comparison as a means of producing categories where the correlation is always adjectival and, more importantly, corporeal.

While bodies produce voices by articulation and engagement of the vocal apparatus – vocal tract, tongue, mouth, nasal cavity, and so on – a voice is never a thing in itself (Eidsheim 2019, p. 3). Voices are constituted by processes of mediation, including but not limited to their physical production. Similarly, voices are not constrained by the production of language in and as speech; nor are language and speech constrained by voiced articulation. Put differently, while (some) bodies can produce a voice, a voice does not necessarily produce a body as something fixed and immediately knowable. What voices do is contingent on a space of relationships constructed by racial, gender, ability, and class-based understandings of listening, *in relation but not constrained to* the vocal apparatus alone. "There is no a priori voice to be known", Nina Sun Eidsheim argues, "prior to the one formulated in response [to the question of who is speaking]" (Eidsheim 2019, p. 6). Moreover, that question might implicate the body producing a voice in a process of subjectification (and by extension, enculturation). This is not to say that the physiological manifestation of a voice depends on this formulation to exist but instead that this relational space is what produces the voice or, in other words, that it is the *listener* who produces the desire to know a voice and, in so doing, creates it as an amalgamation of sound and/within social constructs (Eidsheim 2019).

Similarly, there can be no "a priori" articulation of language through and with the voice that is untainted by an accent; rather, an accent is manifested only at the moment of perception of an utterance (Gergely 2019, p. 145). Accents constantly mutate and are never anchored to a place or the body of the utterer. Rather, they are enmeshed within a larger web of social, cultural, economic, and political practices. Particularly in the case of "border tongues" – that is, the multiple dialects emerging from borders being constantly redrawn and/or crossed – speakers shift accordingly to accommodate the listener, the social situation, the place in which the speech is manifested, or the emotional connections that language may engender (Anzaldúa 2007, p. 77). Speakers of "border tongues" embody the ambiguity of language, accentuating the opacity of certain forms of expression, often remaining unplaceable, elusive, and unmeasurable in detail. Whereas it might still be possible to flag, label, or tag them, such a possibility is more revealing of the systems of flagging, labelling, or tagging (and also of listening) than it is of the speakers themselves.[4] Thus, the ephemerality of accents stands in direct opposition to a reduction to place of origin; if an accent is indicative of social status, class, gender, and ethnicity, it is so with regard not to the *speaker* but to the *listener* producing that judgement. When attempting to erase the listener-observer from the listening moment, we fail to account for the impact of the listener's presence, vision, and bodily engagement – including their listening – on the sonic event. In fact, the discursive erasure of the listener is what sediments the myth that sounds are invested with an ontological origin that not only precedes symbolic orientation but, more importantly, can be disclosed and made known (Thompson 2017). In other words, modern/colonial listening fails to consider that all listening is situated listening (Goh 2017) and that "any meaning that arises is based on [the listener's] own meaning derivation" (Eidsheim 2019, p. 59).

When Machines Listen

While phonograph recordings provided an aural representation of the voice that was subjected to close listening to produce new categories of the human and thus contribute to the development of scientific knowledge, visual translations of sound into other media have rendered "speech spectra particularly amenable to measurement and comparison" (Li and Mills 2019, p. 131). In another historical context – mid-twentieth-century USA – voice recordings, or "vocal portraiture", supplanted earwitness testimony as forensic evidence and inaugurated the representation of the voice as a (proto-)biometric technology. In earlier models of vocal portraiture, listening was assumed to be merely a mechanical process (Günter Tillman in Mills 2010, p. 38); interpretation, accordingly, relied on specific "listening skills" that were aided by the indexical translation of a waveform into another medium. Needless to say, these methods of identification relied not on an alleged objectivity ascribed to the process of reading visual markers but instead on speculative interpretations of what the specific sonic information imprinted on a medium was considered to "expose".

Machine listening is perhaps best understood as emerging at this moment of transition from legibility to detectability, that is, from the descriptive and indexical processing of voice recording to probabilistic value and meaning extraction. Newer models of "voiceprints" did not depict sound waves *per se* but, instead, *representations of their perception* by the inner ear (Mills 2010, p. 38), whereby the inner ear was believed to be "an array of filtering elements" (Potter et al. 1947, p. 10).[5] According to Li and Mills, this transition from *speaker identification* to *speaker recognition*, where "'low-level' acoustic features" replaced the descriptive characteristics of sound (Li and Mills 2019, p. 132) for the evaluation of recorded speech, was grounded in the military and industrial interests of the mid-twentieth century. While modern applications of this model of "machine listening" are not constrained to speech analysis alone, we situate voice biometry as both formative of and as well as remaining within this paradigm of machine listening, in which the coupling of skilled listener-interpreters with a sound wave inscription device established a new paradigm with which to uncover the "unmarked" characteristics of speech.

One such device was the sound spectrograph. Developed by Bell Telephone Laboratories in the 1940s as a "visual hearing" device that could teach "visible speech" to deaf individuals by means of translating sonic representations of speech into words, the technology was then used during the Second World War to satisfy US military interest in "a visual translation of sound" (Potter et al. 1947, pp. 4–5). In a process akin to modern FFT analysis,[6] the spectrograph attempted to mimic the ability of the inner ear to parse multiple frequency bands in isolation, although relying on the different temporality afforded by recorded sound. Instead of employing a large number of filters to perform this operation – which would only later become possible with more computing power – the mechanical spectrograph utilised just one adjustable narrow- and wide-bandpass filter to perform a series of repeated measurements of the same recording, processing, "one band a time, the simple oscillations of a complex wave, and [recording] the intensity variations in each band side by side in an orderly fashion upon a sheet of paper" (Potter et al. 1947, p. 11).

The sound spectrograph employed reduction and repetition to produce a visual representation of *speech, not the speaker*. Its main purpose was to generate transcriptions that could be recognisable by trained translators through pitch, intonation, and timbre. Yet by comparing recordings and clustering recognisable attributes into larger descriptors, spectrograms were also believed to reveal "vocal characteristics imperceptible to the ear", making "accent, gender, and affect visible" and identifying "the unique characteristics of a person's throat, oral cavity, and nasal passage" (Li and Mills 2019, pp. 138–139). As such, spectrograms were believed to emphasise the idiosyncrasies of an individual through their speech, which once again led sound spectrography to inhabit and reinforce the "forensic fantasy of speaker identification" (Li and Mills 2019, p. 137).

Machinic Listening

Today, methods for sound analysis, including speech recognition, still rely extensively on time-frequency representations of sound to derive readable measurements for diverse applications. State-of-the-art implementations include machine learning, where larger databases of speech (*speech corpora*) are processed and analysed to train and calibrate speech recognition, text-to-speech, and speaker identification systems. Much like Wilhelm Dögen's prisoner-of-war archives from the 1920s and 1930s, these databases are thoroughly annotated to highlight characteristics of speech, such as inflection and articulation, and to mark off paraverbal sounds, such as coughing, breathing, and hesitating (de Oliveira and Pedro 2019). Moreover, much like the Bell Labs spectrograph, these recordings are often processed through an array of different filters and discretely analysed to produce visualisations and other articulations designed to satisfy specific industrial, commercial, and geopolitical needs – for example, the enforcement of borders, profit for contractors, and so on. Nonetheless, despite employing significantly larger computing power to parse much larger numbers of data points on an infinitesimal scale, some assumptions remain the same: by measuring the periodicity of a given spectrum (or an *echo*), these analyses are believed to mathematically represent the "vocal tract impulse response, the glottal pulse shape, and the periodic excitation" of speech (Oppenheim and Schafer 2004, p. 97). Furthermore, it is assumed that those representations can be used to derive a person's vocal tract shape to identify their language, accent, or dialect (Dusan and Deng 2000; Ma and Fokoué 2014).

Feature extraction is the fundamental operating principle of machine learning *and* machine listening, and hence constitutive of voice biometry. Louise Amoore defines feature extraction as a set of practices in automated machinic processes which "identify clusters within the data (to read), derive attributes from those clusters to make them recognisable into the future (to resolve), and compare the output of the algorithms with the target output of the analyst (to reason)" (Amoore 2020, p. 43). That they strongly rely on deriving recognisable attributes implies that these methods of attribute-finding are, in themselves, methods of attribution. Indeed, from the moment a recorded voice is inscribed into a system from which it is possible to derive, train, and learn a "lexicon" of features – for example, the mean fundamental frequency, loudness, rate, harmonics – it also begins to inhabit a macro-analysis space of "*perceived* markers" (Li and Mills 2019, p. 138, original emphasis), such as "phonation", "resonation", and "articulation" from which it becomes possible to parse "relevant" information. However, these lexicons are only able to provide information that can be compared against measurements obtained using the same method of analysis, usually discarding those that over- or underfit the prescribed model.

We can thus infer from Amoore's above definition that feature extraction inhabits not only the technical and mathematical space of parsing, filtering, and stratifying frequency data but also that of subjectivity, cognition, and – most importantly – interpretation. Engineering decisions inform which of these perceived markers are claimed to *represent* speech (Mills 2010) to accommodate the features that demonstrate less variability once "appropriate 'normalising relationships' for interpreting the data" can be applied to them (Li and Mills 2019, p. 150). The "feature space" of machine listening, that is, the scope of readable points generated by identification and constitution of clusters of data, from mid-twentieth-century "voiceprints" to today's large datasets, is inherently dependent on filtering out information, both by technological as well as interpretive means, to generate "useful" readings of recorded voices. In other words, the definition of "useful" within datasets is in itself a product of the *intention to use them* in certain ways rather than others (Ahmed 2019).

If, to follow Jonathan Sterne, sound technologies "depend on us delegating our hearing to machines that hear for us" (Sterne 2003, p. 41), it is crucial to ask what forms of hearing are delegated to which machines and for what purpose. This would involve accounting for an entire ecology of the different subjectivities at play, in which different listenings emerge in the construction of a system as already marked by racial, gendered, ability, and labour relationships. Discursively, however, this tends to follow the binary logic of a metaphorical "sensory opposition" between human and machine, defined in terms of subjectivity versus objectivity or bias versus neutrality, so as to render machine listening ancillary to human listening and not the other way around. Nonetheless, despite being framed as distinct (and often separate) techniques or modes of listening, machine listening depends on human listening *at each and every step* of its performance, from the selection and recording of subjects of a certain language to the annotation of databases, the design of evaluation conditions, machine and system calibration, testing, and, ultimately, iterative validation of the distribution of probabilities as truthful or false.

The delegation of listening, that is, the extension of the listener-interpreter's skills to the machine to "help it work" (Sterne 2003, p. 247), exposes the *desire* to produce truth from the recording rather than the ability of the recording itself to produce (or to ever contain) such truth. Nevertheless, when it comes to machine listening and its applications in voice biometry, what Sterne identifies as a "faith in the social function" (Sterne 2003, p. 219) of sound technologies is extended to encompass faith in the ontological conditions of that social function, that is, the ability of machine listening not to *reproduce* but rather to *reveal* relationships that are assumed to be present but imperceptible to human listening. The goal of this framing is to generate a self-reinforcing system in which hypotheses are confirmed in a teleological manner. Thinking of feature extraction in this way, that is, as an "iterative and experimental process in which the humans and machines feel their way toward solutions and resolutions to otherwise indelibly political situations and events" (Amoore 2020, p. 43), further complicates the boundaries between "human" and "machine" listening.

While feature extraction is the underlying mechanism of machine listening, automated analysis of mathematical representations – data, if you will – of sound is nothing but processes of pattern-finding and, more importantly, of generating "statistical distributions of a pattern" (Pasquinelli 2019). As these processes occur in abstract domains that do not involve the physical process of sonic mediation, it follows that machine listening forecloses not only the interpretative, ambiguous, situated, and affective characteristics of listening but also the very *possibility of listening itself*. This would imply, to paraphrase Matteo Pasquinelli's apt provocation (Pasquinelli 2019), that the "naive question" of "Can a machine listen?" should be reformulated as "Can a statistical model listen?". Nevertheless, we believe the issue is more complicated than this. We argue that what defines machine "listening", that is, a process of pattern recognition aimed at uncovering seriality, marking difference, and discarding contingency, is in fact the very same onto- and epistemological framework that defines "human" listening. We do not mean to imply, however, that machines indeed *do listen*; nor do we seek to grant excessive agency to machine listening in relation to (or to the detriment of) human listening. Instead, we emphasise that in this context the issue of "listening", rather than being simply the problematic use of metaphors,[7] might also encompass the transferability of specific orientations towards what "listening" (as a sense) is, and what it produces materially, discursively, and epistemically. This is particularly crucial with voice biometry, in which decision-making processes are delegated to machine listening so as to obscure and prevent human accountability.

The Voice-Language Nexus in Accent Testing

The inclusion of biometric data extraction as an integral part of asylum law secures a regime of practices of identification based on anchoring measurable, corporeal features to identity. However, to the extent that the collection and evaluation of these features are legal devices with which to determine origins (and by extension citizenship), they in effect conflate the physical with the cultural, the flesh with geopolitics, and the bodily with the discursive. In other words, biometry establishes that identity and the body are indivisible, thereby creating an ephemeral "biometric body" which is not only "phenomenologically unknowable" (Liljefors and Lee-Morrison 2015, p. 55) but also "ontologically insecure" (Browne, p. 110). Moreover, biometry denies the subject the possibility of a first-hand experience of identity, for it relies on measurements and operations that are and remain, by design, inaccessible to cognition. In effect, it forges a practice of "identification" – that is, of making the body legible and detectable – that acts as a placeholder for "identity" (the embodied subjectivity and collective lived experience of an individual) (Amoore 2006).

Indeed, German Asylum Law (Asylgesetz, or AsylG) states that the "identity" of an asylum seeker can only be secured by means of identification processes (sec. 16, para.1 s.1. AsylG). This is the case when an asylum seeker is either unable to provide documentation of their origins – passports or other officially recognised papers – or when the case worker suspects the documents presented might be forged. The AsylG lists not only common methods, such as fingerprinting and passport photos, but also voice, or, more precisely, "the spoken word" (*das gesprochene Wort*, sec. 16, para.1 s.3 AsylG), as material evidence for ensuring identification. In Germany, *Sprach- und Textanalyse* (STA) has featured as a practice in asylum cases since 1998 (Lindroos and Kirchner 2018, p. 122), although few cases are heard in court (Lindroos and Kirchner 2018, 125).[8] Nevertheless, the German Federal Office for Migration and Refugees (*Bundesamt für Migration und Flüchtlinge*, henceforth the BAMF) has pioneered a public-private partnership for "IT-supported" asylum processing programmes, including (but not limited to) automating the process of identification through voice analysis (BAMF 2018e).

The BAMF introduced their programme of integrated identity management (IDM-S) with a pilot phase of their "Dialect Recognition Software" (*Dialekterkennungs-*, sometimes *Dialektbestimmungssoftware*), or "Language Biometry" (*Sprachbiometrie*), first tested in Bamberg in August 2017 (BAMF 2018b). At the beginning of 2018, the pilot was extended to other field offices in the country, and, in little more than a year, the BAMF deployed such software in more than 18,000 asylum cases (BAMF 2018g). While STA tends to rely on longer samples, recorded in "conversational protocol" (BAMF 2017), with its software, the BAMF claims to be able to produce "independent, objective, and scalable" (BAMF 2017) proof of origin from only two minutes of "useful" speech (*Nettosprachdauer*). At the time of writing (April 2021), Germany remains probably the only nation-state to have implemented automated, software-led evaluations of language and dialect.

The software functions by matching the speech of the applicant, recorded via a telephone or headset, with a "speech database" (*Sprachdatenbank*) and a "primary language model" (*grundlegendes Sprachmodel*) (BAMF 2018a, pp. 5–6). The results, produced by the software in almost real time, are not decisive but rather present a list of distributed probabilities which are then interpreted by the case worker before a decision is taken (BAMF 2018b). Whenever the software is unable to recognise the language spoken, it returns two different categories: "unknown language" (*unbekannte Sprache*) or "other languages/dialects" (*andere Sprachen/Dialekte*). The language model, initially constrained to the Levantine variation of Arabic,[9] was acquired from the Linguistic Data Consortium at the University of Pennsylvania (BAMF 2018f). A quick search for matching databases at the LDC shows that from the corpora available at the time of the BAMF's implementation, the Levantine Arabic collection was recorded and released in 2005 (BBN Technologies et al. n.d.), while other databases, released after 2018, feature recordings from 2011. When state representatives (*Abgeordneten*), including Ulla Jelpke, Dr. André Hahn, and Gökay Akbulut, enquired into whether the software took into account that "one's language fluency could change over the course of a lifetime" (*dass sich die Sprachkompetenz über die Lebensspanne verändern kann*), the BAMF maintained that language acquisition was "irrelevant" due to the software's focus on "phonetics" (BAMF 2018b, p. 8).

If biometry not only constitutes the body as readable but is also constituted by readings of the body, it follows that biometric technologies produce a body extrinsic to that which is read. This body is constructed (and racialised) precisely at the moment of measurement (Browne 2015, 109) and reified through its connection to other machinic and non-machinic measurements, matches with databases, and insertion into social narratives, all of which, in turn, iteratively affirm and update said narratives and databases for future readings. The BAMF has disclosed little information on the particularities of the language analysis and classification process that its software undertakes. Its insistence on mentioning scientific criteria, such as "phonetics", is intended to validate the workings of the software; however, what it does, instead, is to highlight the limits of identification methods and how they stand at odds with the lived reality of asylum cases. While they do disclose that the margin of error, initially reported to be 20%, has been reduced to 15% in general and to 10% in the specific case of Syrian asylum claimants (BAMF 2017, 2018f), test results often show probabilities so evenly distributed that they become irrelevant to any form of decision-taking (Biselli 2018). In practice, this means applicants risk being not only misidentified by the software but also evaluated by a system that possesses no knowledge of their language.

Nonetheless, the nationwide application of this software has been praised for its apparent "success" (BAMF 2018c), with novel development packages proposed to extend the scope and implementation of IDM-S and MARiS (*Migrations-Asyl-Reintegrationssystem*) to include machine learning and blockchain technology (BAMF 2018e). This reference to machine learning seems to contradict earlier documents in which the BAMF distanced itself from this method (BAMF 2017, p. 3), although some of the criteria displayed on the language report seem to allude to the use of machine learning even in the current iteration of the software.[10] Accurate information on this issue is important, as it would help us constrain the scope of the operativities of machine listening employed by the dialect recognition software and compare it against the BAMF's own reasoning.

The language employed in official documentation, both visible in training documents and disclosed in IFG (*Informationsfreiheitsgesetz*) requests, indicates that the software performs "speech", or more precisely "language", biometry (*Sprachbiometrie*). Consent form D1728, which concerns language analysis and which is given to asylum applicants, states that during the application process "a biometric analysis of the voice does not take place" (*Eine stimmbiometrische Analyse von Ihnen erfolgt dabei nicht*) (BAMF 2020).[11] Nevertheless, the training documents show that, subsequent to the test, two different reports are to be generated: the first, named "Sprachbiometrie

Report", which is automatically generated, presents the distribution of probabilities of spoken dialects as evaluated by the software. The second, to be completed after proofing the result file and after the official hearing has been conducted, asks the case worker to confirm or reject the linguistic claim initially made by the applicant vis-à-vis the test results. This latter document is named "Stimmbiometrie_Auswertung" (BAMF 2018d, p. 10).

The fact that form D1728 explicitly differentiates "language biometry" from "voice biometry" implies that the BAMF regards these two tasks as different and, more importantly, that only one – *Sprachbiometrie* – will be used in the course of the examination. However, the wording within MARiS demonstrates that biometric voice extraction is, indeed, not only performed but distributed: while the software performs "only" the language measurement, a human listener is always responsible for reading, interpreting, listening, and providing a final evaluation explicitly named as a biometric evaluation of the *voice*. Now, even though the BAMF mentions in most official communications that the use of *Sprachbiometrie* is meant only to "aid" case workers in placing the asylum seeker within an ethnic and geopolitical grouping (e.g. BAMF 2017, 2018b, 2018f), what it does, in fact, is to attempt to define "identity" as that which is constrained to mathematically parsed data collected by distinct apparatuses of measurement and thus (allegedly) more "true" than the lived, subjective, and embodied experiences of the targeted individual. The software nevertheless possesses immense power to determine the future of the applicant. Moreover, while relying on this form of "acoustic intelligence" is a means of expediting the identification process and lessening the human workload, it does not necessarily imply a better result for asylum seekers, although the BAMF claims to have no specific records of how many asylum applications have been denied with the aid of biometric language results (BAMF 2018f).

The *Sprachbiometrie* apparatus of the BAMF is, in its appraisal of machine listening, contingent on human listening to produce the body of the asylum seeker as defined by their vocal articulation. This is not to claim that the measuring of acoustic features is an entirely futile exercise. Rather, we simply wish to emphasise that it is far from self-evident that the measurements produced by acoustic feature extraction can – juridically and symbolically – stand for ethnic, geographic, gendered, or racial origins. The selection logic of feature extraction, as well as the interpretation of the results, is always already embedded in a system of categorisations constructed by modern/colonial and therefore racialised understandings of voice, speech, language, and accent. That is, the software cannot produce anything other than the categories it was designed to produce, insofar as they give "the power to confer on the analyst the power to speak in their name" (Amoore 2020, p. 54). This power is further manifested in and through juridical documentation, policy papers, asylum laws, and press statements; these documents confirm not only that the software does what it intends to do but also that this doing participates in a longer chain of decisions whose ultimate intent is to prove an irreducible connection between body and accent and to tie that accent irreducibly to citizenship. Simply put, it is not that the accent with which someone seems to speak determines their right to asylum, but, instead, that the biometric extraction of their accent is already determined by the possibility of deportation.

Such a field of possibilities is created, on the one hand, by careful wording choices on the juridical side, and, on the other hand, by a concern with technical terminologies on the discursive side, thus evincing how the BAMF establishes its own "conditions of intelligibility" (Butler 1997, p. 134). In other words, it constantly negotiates the boundaries of terminologies such as "*Sprach-*" and "*Stimmbiometrie*" to remain ambiguous and escape "the terms of legibility that it occasions" (Butler 1997, p. 134), where this legibility is both its own institutional ethos (i.e. deportation) as well as the legibility of the asylum seeker's body before the State. Moreover, as it is probably clear by now, these conditions of legibility emerge discursively through, but are not necessarily executed by, acts of listening; neither the case worker, attorney, interpreter, nor asylum seeker actually *listens to the software listening to accents*. This "listening" remains obscure and unavailable to human cognition, transversing technoscientific and juridical language when translated into statistical distributions and log likelihood ratios. When listening does take place, it occurs at a different moment (the hearing), is performed by a different actor (the case worker), and is employed to complete (confirming or rejecting) a claim made not by the asylum seeker but by the algorithm on their behalf.

Conclusion

Rather than residing exclusively in the body, voice, speech, and accent are phenomena that emerge and are constantly renegotiated between listener and sound. Nonetheless, when it comes to the management of bodies, it is assumed that by applying the technical apparatuses of sound visualisation, one can identify these phenomena, which are assumed to be produced uniformly across bodies that share similar *citizenship* and therefore similar bodily features. In other words, it ascribes *racial value* to a presumed "innate" capacity of the body not only to produce a specific voice but also to produce specific articulations of speech (i.e., accents) that can be normalized, tied to place and citizenship, and made immediately available for extraction, measurement, and evaluation. The example provided here sits alongside other critical scholarship that demonstrates how modern/colonial listening practices establish a teleological relationship between voice and body. It is assumed, on the one hand, that voice produces a knowable body but, on the other hand, that what counts as voice is intrinsically dependent on *who* counts as a subject, or at least who can be made to be that way through (colonial) enculturation.

Voice biometry iteratively co-constructs bodies and identities as an articulation between nation-states and civil society, borders and their industries, national identities and social status, and human and machine listening. It reaffirms itself "through a selective process in which the terms of legible and intelligible subjecthood are regulated" (Butler 1997, p. 41). Identification techniques, in particular those that utilise voice, are therefore *performative practices*; that is, they are invested in producing "social effects", imbued with power "not only to regulate bodies, but to form them as well" (Butler 1997, pp. 158–159). Machine listening, in that regard, does not necessarily concern only the delegation of hearing and the interpretation of visual markers to a machine; instead it also represents an orientation towards inscribing interpretative value to what automated readings of extracted features produce and the insertion of said interpretative value into a techno-scientific understanding of voice as intrinsic to body and body as intrinsic to a (produced) identity. That this happens on the level of "machine" listening while requiring completion by a human listener only confirms that the biometric moment occurs in neither moment alone, but rather at the in-between.

In the case presented here of so-called "accent recognition" software, the testing moment, in which two minutes of speech are recorded, does not produce a story, struggle, truth, or identity; instead, it produces a readout of the conditions of probability measured by the software and the software alone, which in turn constrain the scope of judgement allocated to the case worker. Such incongruence highlights how the software itself is designed to yield assessments of accents that are connected by geographical, ethnic, and cultural proximity, disclosing how "the accent exists in the ear of the listener [and is] linked to the space of the imagined community" (Gergely 2019, p. 147). This is clear, for instance, when implying that such software is capable of recognising the difference between an unknown language and an unknown dialect, while at the same time rooting language and dialect in a body which is assumed to be equal to a whole community of other bodies speaking the same languages or dialects.

If accent or dialect testing is closely related to other biometric practices that produce extrinsic bodies, this specific biometric body is produced as an extrinsic, sonic body that is, paradoxically, rendered silent by the automated measurement of its own language and voice. Accent and dialect testing are colonial instruments constantly attempting to "tame a wild tongue" (Anzaldúa 2007), or at least what their instruments consider a "wild tongue" to be. They discard the heterogeneity of articulation, wording, and code-switching in favour of fixed, reproducible narratives. However, in their simplistic reduction to "phonetics", "acoustics", or a combination of both, they fail to consider the material conditions that shape and are shaped by the journeys that occur before and beyond the moment of crossing a border. The question "where are you from?" has surely been asked of an asylum seeker many times (Khosravi 2020); the answer, it seems, neither the BAMF, the software, nor the caseworker is able to listen.

Acknowledgements

This chapter was written partially thanks to a KONE Foundation Postdoctoral Fellowship in the Arts at the Helsinki Collegium for Advanced Studies (HCAS) in 2021. It has been revised and copyedited for grammar by Matthew Billington at the Language Center, University of Helsinki.

Notes

1 For instance, otoacoustic emissions (cf. Swabey et al. 2004).
2 It is ironic that European forms of rationality and measurement, especially with regard to the fixation of tone and rhythm, and thus the emergence of the polyphonic notation of rhythm around 1300, are only possible via cultural influences from Arabia via Spain. The desire to "master" otherwise oral-based practices and arts like music by storage and standardisation (by the Pope/Church slightly earlier throughout eighth century) was the motivation for developing media based on power.
3 Original quote in German: "zum Zwecke des mittelbaren Verständnisses des völkischen Kultur- und Geisteslebens".
4 Meaning that the practices of labelling and tagging are revealed at this moment of reductionist attribute-finding. For more on this discussion, see Amoore 2020, especially Chapter 3.
5 As both Mills and Sterne rightly note, this was not only based on but also helped establish ableist understandings of speech, sound, perception, and deafness (Mills 2010). Sterne also notes Alexander Graham Bell's ableism towards deaf culture in approaching "visible speech" with a "corrective" impetus, "seeking to eradicate deaf culture altogether" (Sterne 2003, p. 40).
6 FFT does not employ bandpass filtering but instead relies on short-term spectra (around 10 milliseconds) to decompose a complex waveform.
7 For more on metaphors in machine learning, see Ganesh (2020).
8 While STA is a specific German term, it does strongly resemble LADO, or Language Analysis for Determination of Origin. Both employ linguists and often trained "native speaker experts" working in tandem to listen and analyse asylum hearings and interviews and in so doing produce a "flawless, clear and unambiguous" (Ammer et al. 2013, p. 295) report on the applicant's mother tongue, their origins, and therefore their citizenship. Despite claims of standing "aloof from ideology" (Hoskin 2018, p. 27), LADO (and by extension STA) has been strongly criticized for its lack of transparency and accuracy; its reports have led to numerous deportations and, in some cases, left asylum seekers in bureaucratic and political limbo (Hoskin 2018; Ammer et al. 2013).
9 The software was implemented in response to the influx of migrants coming from the Levant region – Syria, Lebanon, Jordan, and Palestine, in 2015.
10 These are, for instance, the "segmentation score", "gender score", and "log likelihood ratio". Moreover, there is no mention or evidence of a transcription of the

voice sample being added to the case file, which seems to point towards a hybrid algorithmic model based on both phonotactic and acoustic approaches (Brown 2016; Ma and Fokoué 2014; Biadsy et al. 2010). While this does not necessarily implicate machine learning, state-of-the-art research on dialect recognition systems (including those created by the company that developed the software for the BAMF, Nuance Communications) has constantly employed Artificial Neural Networks in hybrid models such as these for improved performance (see, e.g., Kanervisto et al. 2017). Whether the BAMF uses this specific model remains in the realm of speculation.

11 According to the BAMF, Form D1728 is issued only in German. (BAMF 2020).

References

Ahmed, Sara, 2019. *What's the Use?: On the Uses of Use*. Duke University Press, Durham.

Ammer, M., Busch, B., Dorn, N., Rienzner, M., Santner-Wolfartsberger, A., Schicho, W., Seidlhofer, B. and Spitzl, K., 2013. Ein umstrittenes Beweismittel. *juridikum* 2013, 281–297.

Amoore, Louise, 2006. Biometric borders: Governing mobilities in the war on terror. *Political Geography* 25, 336–351. https://doi.org/10.1016/j.polgeo.2006.02.001

Amoore, Louise, 2020. *Cloud Ethics: Algorithms and the Attributes of Ourselves and Others*. Duke University Press, Durham.

Anzaldúa, Gloria, 2007. *Borderlands/La Frontera: The New Mestiza*, Third Edition, ed. Aunt Lute Books, San Francisco.

BBN Technologies, Makhoul, John, Zawaydeh, Bushra, and FrederickChoi, David Stallard, n.d. BBN/AUB DARPA Babylon Levantine Arabic Speech and transcripts. https://doi.org/10.35111/J1TN-V351

Biadsy, F., Soltau, H., Mangu, L., Navratil, J. and Hirschberg, J., 2010. Discriminative phonotactics for dialect recognition using context-dependent phone classifiers. In: *The Speaker and Language Recognition Workshop. Presented at the Odyssey 2010*, Brno, Czech Republic, pp. 263–270.

Biselli, Anna, 2018. Das BAMF will seine Probleme mit Technik lösen – und macht alles noch schlimmer [WWW Document]. *netzpolitik.org*. https://netzpolitik.org/2018/das-bamf-will-seine-probleme-mit-technik-loesen-und-macht-alles-noch-schlimmer/

Brown, Georgina, 2016. *Automatic Accent Recognition Systems and the Effects of Data on Performance*. Presented at the Odyssey 2016, pp. 94–100. https://doi.org/10.21437/Odyssey.2016-14

Browne, Simone, 2015. *Dark Matters: On the Surveillance of Blackness*. Durham, Duke University Press.

Bundesamt für Migration und Flüchtlinge, 2017. Drucksache 19/190: Einsatz von Akzenterkennungssoftware durch das Bundesamt für Migration und Flüchtlinge [WWW Document]. http://dipbt.bundestag.de/extrakt/ba/WP19/2303/230303.html (accessed 3.31.21).

Bundesamt für Migration und Flüchtlinge, 2018a. Integriertes Identitätsmanagement IDM-S: Benutzerhandbuch "Sprachbiometrie". https://fragdenstaat.de/anfrage/foliensatze-und-interpretationshilfen-zu-sprachanalyse/ (accessed 3.31.21).

Bundesamt für Migration und Flüchtlinge, 2018b. Drucksache 19/1663: Einsatz von Spracherkennungssoftware durch das Bundesamt für Migration und Flüchtlinge [WWW Document]. https://kleineanfragen.de/bundestag/19/1663-einsatz-von-spracherkennungssoftware-durch-das-bundesamt-fuer-migration-und-fluechtlinge (accessed 3.31.21).

Bundesamt für Migration und Flüchtlinge, 2018c. eGovernment Preisverleihung: 1. Platz. *Infothek des BAMF*. https://www.bamf.de/SharedDocs/Meldungen/DE/2018/20180621-am-egovernment.html (accessed 3.31.21).

Bundesamt für Migration und Flüchtlinge, 2018d. Dienstanweisungen zum Umgang mit der Handyauswertung [WWW Document]. https://fragdenstaat.de/anfrage/dienstanweisungen-zum-umgang-mit-der-handyauswertung/(accessed 3.31.21).

Bundesamt für Migration und Flüchtlinge, 2018e. Digitalisierungsagenda 2020 [WWW Document]. *Infothek des BAMF*. https://www.bamf.de/SharedDocs/Anlagen/DE/Digitalisierung/broschuere-digitalisierungsagenda-2020.html?nn=282388 (accessed 3.31.21).

Bundesamt für Migration und Flüchtlinge, 2018f. Drucksache 19/6647: Einsatz von IT-Assistenzsystemen im Bundesamt für Migration und Flüchtlinge [WWW Document]. https://kleineanfragen.de/bundestag/19/6647-einsatz-von-it-assistenzsystemen-im-bundesamt-fuer-migration-und-fluechtlinge (accessed 3.31.21).

Bundesamt für Migration und Flüchtlinge, 2018g. Sprachbiometrie-Software [WWW Document]. *FragDenStaat.de*. https://fragdenstaat.de/anfrage/sprachbiometrie-software/ (accessed 3.31.21).

Bundesamt für Migration und Flüchtlinge, 2020. D1728 – Information zur Abgabe einer Sprechprobe für die Sprachbiometrie. https://fragdenstaat.de/anfrage/belehrung-von-antragstellenden-sprachbiometrie-dias-namenstranskription-tralita/ (accessed 3.31.21).

Butler, Judith, 1997. *Excitable Speech: A Politics of the Performative*. Routledge, New York.

Dögen, Wilhelm, 1925. *Unter fremden Völkern*. Otto Stollberg, Berlin.

Dusan, S. and Deng, L., 2000. Recovering vocal tract shapes from Mfcc parameters. In: *Proc. ICSLP*. pp. 3087–3090.

Eidsheim, Nina Sun, 2019. *The Race of Sound: Listening, Timbre, and Vocality in African American Music, Refiguring American Music*. Duke University Press, Durham.

Ganesh, Maya Indira, 2020. The ironies of autonomy. *Humanities and Social Sciences Communications* 7, 157. https://doi.org/10.1057/s41599-020-00646-0

Gergely, Gabór, 2019. Sonority, difference and the Schwarzenegger star body. *Film-Philosophy* 23, 137–158. https://doi.org/10.3366/film.2019.0106

Goh, Annie, 2017. Sounding situated knowledges: Echo in Archaeoacoustics. *Parallax* 23, 283–304. https://doi.org/10.1080/13534645.2017.1339968

Hoskin, J., 2018. Native speaker non-linguists in LADO: An insider perspective. In: Nick, I.M., and Cambier-Langeveld, T. (Eds.), *Forensic Linguistics: Asylum-Seekers, Refugees and Immigrants*. Vernon Press, Wilmington, Delaware, pp. 23–40.

Kanervisto, A., Vestman, V., Sahidullah, M., Hautamaki, V. and Kinnunen, T., 2017. Effects of gender information in text-independent and text-dependent speaker verification. In: *2017 IEEE International Conference on Acoustics, Speech and Signal Processing (ICASSP). Presented at the 2017 IEEE International Conference on Acoustics, Speech and Signal Processing (ICASSP)*, IEEE, New Orleans, LA, pp. 5360–5364. https://doi.org/10.1109/ICASSP.2017.7953180

Keshavarz, Mahmoud, 2019. *The Design Politics of the Passport: Materiality, Immobility, and Dissent*, First published. ed. Bloomsbury Visual Arts, London New York Oxford New Delhi Sydney.

Khosravi, Shahram, 2020. *Bordered Imagination*. Crisis Magazine.

Lautarchiv der Humboldt-Universität zu Berlin, n.d. *Bestand "Phonographische Kommission"*. Humboldt-Universität zu Berlin, Berlin.

Li, Xiaochang and Mills, Mara, 2019. Vocal features: From voice identification to speech recognition by machine. *Technology and Culture* 60, S129–S160. https://doi.org/10.1353/tech.2019.0066

Liljefors, M. and Lee-Morrison, L., 2015. Mapped bodies notes on the use of biometrics in geopolitical contexts. In: Michelsen, A. and Tygstrup, F. (Eds.), *Socioaesthetics: Ambience - Imaginary, Social and Critical Theory: A Critical Horizons Book Series*. Brill, Leiden; Boston, pp. 53–72.

Lindroos, E. and Kirchner, S., 2018. Language analysis for the determination of origins (LADO) in German asylum law. In: Nick, I.M. and Cambier-Langeveld, T. (Eds.), *Forensic Linguistics: Asylum-Seekers, Refugees and Immigrants*. Vernon Press, Wilmington, Delaware, pp. 121–132.

Ma, Z. and Fokoué, E., 2014. A comparison of classifiers in performing speaker accent recognition using MFCCs. *Open Journal of Statistics* 04, 258. https://doi.org/10.4236/ojs.2014.44025

Mills, Mara, 2010. Deaf Jam: From inscription to reproduction to information. *Social Text* 28, 35–58. https://doi.org/10.1215/01642472-2009-059

Murray, Heather, 2007. Monstrous play in negative spaces: Illegible bodies and the cultural construction of biometric technology. *The Communication Review* 10, 347–365. https://doi.org/10.1080/10714420701715415

Ochoa, Gautier and Maria, Ana, 2014. *Aurality: Listening and Knowledge in Nineteenth-Century Colombia*. Duke University Press, Durham.

de Oliveira, Vieira and Pedro, J.S., 2019. On the endless infrastructural reach of a phoneme. *Transmediale Journal*.

de Oliveira, Vieira and Pedro, J.S., 2020. Dealing with disaster: Notes towards a decolonizing, aesthetico-relational sound art. In: Groth, P.S.K. and Schulze, P.H. (Eds.), *The Bloomsbury Handbook of Sound Art*. Bloomsbury Academic, New York, pp. 71–88.

Oppenheim, A.V. and Schafer, R.W., 2004. From frequency to quefrency: a history of the cepstrum. *IEEE Signal Processing Magazine* 21, 95–106. https://doi.org/10.1109/MSP.2004.1328092

Pasquinelli, Matteo, 2019. How a machine learns and fails – A grammar of error for artificial intelligence. *SPHERES – Journal for Digital Cultures* 5.

Potter, R.K., Green, H.C. and Kopp, G.A., 1947. *Visible Speech*. Van Nostrand, New York.

Robinson, Dylan, 2020. *Hungry Listening: Resonant Theory for Indigenous Sound Studies*. University of Minnesota Press, Minneapolis.

Sterne, Jonathan, 2003. *The Audible Past: Cultural Origins of Sound Reproduction*. Duke University Press, Durham.

Swabey, M.A., Beeby, S.P., Brown, A.D., and Chad, J.E., 2004. Using otoacoustic emissions as a biometric. In: Zhang, D. and Jain, A.K. (Eds.), *Biometric Authentication*, Lecture Notes in Computer Science. Springer, Berlin, Heidelberg, pp. 600–606. https://doi.org/10.1007/978-3-540-25948-0_82

Thompson, Marie, 2017. Whiteness and the ontological turn in sound studies. *Parallax* 23, 266–282. https://doi.org/10.1080/13534645.2017.1339967

5

Speaking Thr-o-ugh

Putting It Out, Speaking, Listening, Disrupting?

Laura Mello, Mariana Carvalho, Nirah Pomar, and Elif Gülin Soğuksu

Introduction

(Laura Mello)

Su cuerpo es una bocacalle. (Anzaldúa, 2005)[1]
[her body is a "street-mouth"]

A woman is standing on the pavement at Kottbusser Tor in Berlin. People pass by, some look at her, some just move their eyes, some interact with her. The woman has a loudspeaker in her mouth, pointing outwards, so that passersby can hear a female* voice laughing, loud and long. On the other side of the world, in the historic city centre of Florianópolis, Brazil, a woman has a loudspeaker inside her mask – where a filter should be (it's the pandemic) – and it is visible, pointing outwards as well. From the loudspeaker, one can hear another woman's voice, repeating over and over again that she is tired. In Istıklal Street in Istanbul, a woman has a loudspeaker in her mouth and is walking around, the same familiar female* voice is laughing, we already know her from Berlin and Florianópolis. The woman's name is Elif Gülin Soğuksu. She is in Istanbul, Mariana Carvalho is in Berlin, and Nirah Pomar is in Florianópolis, and they are performing *speaking thr-o-ugh*,[2] an intervention developed during a residency, as part of the *Somarumor – International Encounter of Sound Art* in association with the *Topia Sound Art Festival.*

In this essay, we want to recall the process we went through.

DOI: 10.4324/9781003194453-6

Dystopie <-> Topia <-> Speaking Thr-o-ugh

The process of *speaking thr-o-ugh* started in mid-2021 with Zoom sessions with the group members between Berlin, Istanbul, and Florianópolis. The sessions involved myself, a Brazilian living in Germany, who had just come from a residency in Tarabya, Istanbul (where I lived from June to September 2021); Mariana Carvalho, another Brazilian artist living in Berlin; Nirah Pomar, an artist and music therapist who lives in Florianópolis; and Elif Gülin Soğuksu, a Turkish artist who I met through my colleagues from the collective Errant Sound in Berlin, Elif was in Berlin for the first time, doing a residency at Errant Sound.

At that time, I was also involved in the organisation of the *Topia Sound Art Festival* (November 2021). Vanessa De Michelis and I, the other half of the organising team, were pretty much driven by the forum *Listening as a tool to blow out the bubble*,[3] which we organised in 2020, as part of *Dystopie Sound Art Festival*. Back in 2020, when the curators Alexandre Fenerich, Lilian Campesato, and I started to decide on the structure of that forum, we wanted to stress the borders and cultural hegemonies inherent to some definitions of "Sound Art" or "Klangkunst".

The Forum *Listening as a Tool to Blow Out the Bubble*, together with the *Showcase Brazilian Sound Art*, turned out to be unique encounters, a real blast in time, space, and diversity of approaches. Such events wouldn't have been impossible in person, even outside of the pandemic, as travel within Brazil means very long distances and expensive plane tickets. At the same time, physical distance, combined with digital proximity between the participants, brought together different groups of people beyond the usual academic connections.

The 12 works that were part of the *Showcase of Brazilian Sound Art*[4] not only came from artists of diverse origins, but also represented diversity in the way we look at sound art, or better, in the way we listen to sound art. Thus, while organising *Topia 2021*, considering the act of listening as a tool to renew the connection to the outside world and to open a path for new social perspectives, it still significantly resonated in us not only as curators, but also in myself as a mentor on the residence programme of *Somarumor*.

During the first round-table Zoom session at the *Topia Sound Art Festival*, the guest panellists Janete El Haouli and Ximena Alarcón spoke about the role that migrant experiences played in their artistic practices. Ximena spoke in particular about Janete and José Augusto Mannis´s work *Memories of Zahrah*,[5] a radio piece about the life of Janete´s mother. Zahrah moved from Lebanon to Brazil as a young woman accompanying her husband, in one of the big immigration waves in the late 1940s. For Janete, her mother´s voice, recorded on tape one month before her death, is much more than an information container; it is an affect channel. Listening to Zahrah reporting on her life, first in Lebanon, the loss of her father at the age of eight, boarding school, immigration, building a new home in a foreign country, all her expectations, followed by frustrations and successes, a whole life can be heard in the sound of her voice.

But it was not a voice; instead, the loss of it that brought Janete to this perception of the voice. She had taken care of her father, who had lain in bed for nine years, and after having a tracheotomy, his voice could not be heard anymore. Janete was at the time (1979) taking classes with Hans Joachim Kollreuter, the German composer who had just arrived in Brazil after finishing at the Goethe Institute in India. He introduced Janete to Indian traditional music, and later she got to know the work of Demetrius Stratos, an Egyptian-born Greek composer, who was deep into artistic research on the disembodied voice, exploring the voice as sound in all its potentiality, materiality, and physicality.

Janete came across the cover of a Demetrio Stratus album during the time her father was ill, and this left a shocking impression on her: it was the album *Cantare la voce* from 1978, where the illustration on the album cover shows a man with a mouth in his throat. Janete directly connected the cover image to her father´s tracheotomy, and she realised she could not remember his voice anymore. She started missing his voice, and, with it, the culture behind it. For her, the moment had come to start connecting again with her ancestors' culture. Missing her father´s voice brought her to look for it in the story that lived in his silent living being. In the discussion, Ximena was struck by these moments in Janete's life, where she spoke about voices being lost. In El Haouli/Mannis' piece, Janete's mother Zahrah speaks, in a tiny soft voice, about losing her voice whilst working as a teacher on a day when she was teaching a class of 40 children. Her husband, also a Lebanese immigrant, lived for nine years with a tracheotomy, and his family also forgot how he used to speak. Here, Ximena Alarcón noticed the symbolic materialisation of the nomadic voice that loses territory and asked, "where are we when our voices lose territory, or when we lose the voice?" Alarcón also indicates the connection between the loss of one's voice – or the loss of the right to speak and be heard – and the caring for children.

Recent literature on migration has turned towards a focus on female* experiences, using, for example, reports from personal letters, where it's often reported that women accept their husband's decisions and emigrate with them, even if this is not their own choice (Fend, 2020). Janete´s family's experience resonates with this, as her father went to Brazil after being told by his brother, already living there, that there were big opportunities around the Cianorte Coffee Company in Paraná State. It wasn´t Janete´s mother, Zarah´s, wish to leave her life in Lebanon, but she moved to Brazil with her husband. At the end of her life, in the work of El Haouli/Mannis, this is what we hear when listening to her soft voice in the performance.

Urban Voices in the Distance

So, back to our performances in Berlin, Florianópolis, and Istanbul: a single female* puts herself on the street, she transports several other women*´s voices back and forth around the earth from and to two continents. Here, the voices migrate without their bodies. Each of the performers stands for other women in a symbolic way. There are no Elif, Mariana, and Nirah anymore. On one side, they stand for a specific female* identity, but are also individual cultural entities, who have carried migration at some point in their life histories.

As mentioned earlier, it was 2021, and the global pandemic was still keeping events online, so WhatsApp and Zoom were our usual communication tools. Apart from the negative effect on the social side, the possibility of working together from far apart was at the time already intensively integrated into working processes, becoming a kind of aesthetic frame. Instead of considering distance a hurdle, we decided to focus on the idea of the space between three women in three different countries. Thus, our group came together virtually for the first time after the residency process started.

From my notes from the first meeting with Elif, Mariana, and Nirah on 19 July 2021, I find remarks such as "urban daily life", "cosmoaudition – non-hegemonic". Heteroglossia, meaning multiple voices synthesised in one text, or an open dialogue, rather than one common language (Haraway, 1985), I think about Bakhtin, carnival, and the urban space. Urban intervention, the urban as a place where "centre and periphery" meet. Cosmoaudition is a cognitive system of perceiving the world through sound and was one of the key themes of *Somarumor*. Looking at the final results of the residency, it becomes clear that the idea of urban intervention framed the work, allowing us to break up with the pandemic restrictions and to create a situation that could be performed in the same way in three countries, almost on the same day (only Elif went to the street two days later).

Even at this point of the project, there were already some similarities between our artistic approaches: both me and Mariana had already been using loudspeakers as extensions of our bodies, for example. For this project, loudspeakers placed on the artists' mouths, or used as masks, established connections and collisions across the three countries and two continents, between the individual and the collective, the private and the public. Loud or quiet speakers, the title *speaking thr-o-ugh* arose, the "o" as an open mouth, and the "ugh" for what is being put out. Elif, Mariana, and Nirah stand for, present, represent, and at the same time amplify the more than 40 female* voices presented in the project. The voices migrated, got attention in a different place, were rejected or neglected, or even threatened by the police.

The voices heard coming from the loudspeakers in the interventions were collected during a call the artists made, inviting female* identified contributors to answer the question "What do you want to put out?". The samples are audible snapshots from the lives of people who live as female*s somewhere in the world. Female* identifying persons "put it out" what was mostly moving them at the moment in Russian, Spanish, Portuguese, English, and Turkish, but also laughing and yelling. The pool of answers, sent mostly as WhatsApp audio messages, was edited and given a final presentation by the artists. From this final mixed edition, each artist would decide which sample fitted the situation created during the urban intervention in each city.

Speaking Through Speaking Thr-o-ugh

The text below consists of a collection of nonlinear written and transcribed (from audio) conversations between myself (Laura Mello), Elif Soğuksu, Mariana Carvalho, and Nirah Pomar, through which we try to recall how each of us related to the processes of the project and our impressions and thoughts about it.

Mariana: Working with multiple voices forms an intricate weaving, when recording, womxn* we didn't listen to each other's audio messages until we put them in conversation. I'm thinking about those dialogues, about the relationality that is built upon the simple fact of having a listener: saying and waiting for the answer to come. Voicing implies a listener...It brings us to the conversation as a method, that Lílian Campesato and Valéria Bonafé touch on so well:

> Além da vocalização, de tornar audível um pensamento, é necessária a presença de um outro para que uma escuta de si se efetive. Nesse sentido há uma projeção indispensável. Há que se colocar no lugar do outro para se escutar. A alteridade potencializada pela conversa se efetiva muitas vezes no ambiente do compartilhamento de intimidades. A conversa se abre assim a confidências, a essa comunicação particular e frequentemente oral de uma intimidade.
>
> *(Bonafé & Campesato, 2019, p. 38)*

> [Besides vocalisation, making a thought audible, the presence of an-other is needed for the realisation of a listening of oneself. In that way, there is an indispensable projection. In order to listen, putting oneself in the place of the other is necessary. Many times, the otherness, strengthened by the conversation, takes place in the ambient of sharing intimacies. Conversation, therefore, opens up to sharing something personal, to this particular and often oral communication of intimacy.]

As Bonafé and Campesato note, bringing to mind the work of Margareth Rago, it's about sharing intimacies and what comes out of those conversational moments, written or not, is oriented by a listener. I guess, besides the conversation that is formed when performing multiple voices on site, this text is interwoven around the above idea. As explored before, it is a non-linear conversation. It contains fragments of each of us writing, then alternated and organised in topics; transcriptions of a video call between us, as well as transcriptions of some of the shared voice recordings. Not necessarily dialogic, they propose different positions in relation to the work. We write and speak about it, but we also try to let it speak through the different implied voices, or through the agency of the voices in this text itself. Between oral and written, we've been experimenting with juxtaposing and re-orienting. In this attempt of speaking through *speaking thr-o-ugh*, maybe a way we can be offered for multiple listenings of the work and its process.

Voice(S)

> Every sound we make is a bit of autobiography. It has a totally private interior yet its trajectory is public. A piece of inside projected to the outside. The censorship of such projections is a task of patriarchal culture that (as we have seen) divides humanity into two species: those who can censor themselves and those who cannot.
>
> *(Carson, 1995)*

Our Voices

Nirah: I am a forty-year-old white female*, I have lived in Florianópolis, an island in the south of Brazil, for thirty years: where I grew up, where I study and work, my son was born here twenty years ago. I have studied the voice as an instrument for almost fifteen years – as a musical instrument, as an instrument for expressing the soul, as an instrument for communicating pain and desires, as an instrument for healing. The voice is the instrument that is always with us and yet it is not always available: the voice fails in moments of fear, tension or strong emotion. Often, the voice is stilled, silenced, repressed, smothered – especially if you are a woman.

Elif: The notion of voice has always been something that interested me. A sound-source that I can shape and sculpt in the easiest way, a morpho-plastic instrument through which I can express myself more effectively. Since my childhood, I have imitated the people and the soundscape surrounding me through my voice, transformed and moulded them as I listened; I have changed my identity through my own voice and mutated in the identities of the sounds around me. As well as this, I have silenced my own voice, my own thoughts and sometimes haven't lent my ear to my inner voice(s), haven't shouted or raised my own hesitant murmurings when I should have done.

I have lived in Istanbul for my whole life, for 23 years, until the summer of 2021 when I lived in Berlin for 4 months. For the first time in my life, I went abroad and lived in the multicultural environment of Berlin. I encountered many different languages spoken in a single city; I encountered divergent identities. Furthermore, most radically, I had to convey my thoughts and words with a foreign language in an unfamiliar sounding space and society. I was fascinated by the atmosphere of the city and its vividness, rebellion and freedom; of the residents who were shouting their identities and discrepancies, manifesting themselves freely. I was eager to do the same, to shout, to have the courage to express myself freely. Nirah, Mariana and I, we were all engaged with that ambition to exist, to provoke.

Mariana: I moved from São Paulo – where I'm from – to Berlin in 2020, and since then I've been constantly putting my body into this country through endless clash and adaptation. Bodies, space, language, weather, the city, its silence, every single detail that is radically different from my place of origin. I find myself in reorientation inhabiting the intermediary spaces of borderlands – to evoke Sara Ahmed's Orientations Matter (Ahmed, 2010) and Gloria Anzaldúa's work around the mestiza (Anzaldúa, 2005). I am dis-placed and re-placed; white privileged middle-class cis young woman in Brazil; Latin-American (still white) immigrant in Europe, here not so Western as we consider ourselves in Brazil. Relocating, repositioning, different perspectives. Placing oneself in space in relation to other bodies is a winding task, even more intensified in a shifting pandemic world.

Nirah: It is from this place, as a woman, a single mother, that I have often silenced my voice and my desires before society. It took me a long time to start listening to myself and to find in myself the necessary courage to say what I think and feel. Perhaps this is the point where I most identified with the experience I had during this project: *speaking thr-o-ugh* allowed me to get in touch with the stories of other women who live in situations very similar to everything I have lived.

Multiple Female Voices*

Nirah: We received messages in multiple languages, including jokes, screams and laughter. Some messages reported abuse, others spoke of hope, others just narrated everyday situations. Some voices, sometimes barely audible, seemed to scream for help. What all voices had in common was the desire to be heard: women tired of being afraid of being women, in search of freedom to exist!

Mariana: As people with mouths, sounding like other mouthed-people, we received a lot of strong testimonies. Not only watchwords, but reflections on life, society and even stories of sexual abuse. Screams and laughter would convey non-verbal intensities, and I felt that in each testimony there was a collective urge. Putting out those voices seemed so necessary as it was a hard task; I felt we had a lot of responsibility in dealing with them and making them audible. Those people trusted us and shared delicate stories, giving much personal content. I felt a radical empathy. Each voice is more than one; your voice is mine is ours.

Nirah: "This body is mine", sings one of the women. It may sound crazy, but women often don't own their own bodies. So, we have to scream the obvious: this body is mine!

"It's very difficult to be a woman in this structurally sexist society, but I never thought it would be stratospherically worse to be a mother in that same society", says another woman. "I'm tired, I'm tired, I'm tired", said another one, letting off steam. But this one seems to be the deepest:

As a child, I was abused several times. I once spoke in a theatre exercise about the first time I saw a dick. The first time I saw a dick, it was a fight with him, you know? Because they lowered my shorts, opened my mouth and pushed it inside. I was about five years old. He was an uncle of mine, an idiot asshole, an asshole, a person that everyone in the family likes, that everyone in the family takes care of because he's the youngest brother, idiot. He must have forgotten. He acts as if nothing has happened, but I haven't forgotten. People never forget. After him came a cousin, another cousin, another cousin, another cousin, a neighbour, my grandmother's boyfriend. And last, and not worse and not better, but maybe just more painful, my father. It started just as a joke, it's always like a joke, right? I must have been six, seven. I don't know how old I was, it lasted so long! But he was smart because he didn't go all the way, because he knew that if he killed it was the end, the deal would stink, but he went long enough. Long enough to kill all the affection, all the affection and all the attention that one day I already feel for the one I called a hero. Everyone likes him, everyone has always liked him. He was always the funniest of the brothers, the nicest square, the coolest. My friends used to say "Gee, I wish I had a father like yours" – don't! He died. I swear I didn't wish him dead, I guess. I didn't say goodbye; I had nothing to say. I mean, I had. But would everyone be able to hear it? Would everyone be able to stand to hear and know that the hero was always the villain? Will it be? I have done more for him in death than he has done for me in life. Curious. The teaching he left me, the biggest teaching he left me is the kind of father I don't want to give to my future children. I could have screwed up, I could have gone wrong. I could have failed. I could… have been a mistake. I don't know how I didn't freak out. I don't know how I wasn't traumatised. Maybe I'm traumatised. Maybe I need therapy, psychoanalysis, and a psychologist. I never said… People know… If everyone knows… I'm strong or I play strong, I don't know. Surviving is just about having life, about a beating heart, about having a functioning, not surviving, about keeping your mind, about keeping your mind! Even when everything around you seems to fall apart… I'm alive!

Elif:
(translated from the Turkish original)

Female* voice 1

I specifically demand freedom for women in education, employment and representation. I want to be able to see that women are also leaders, contribute to peace, determine their own future, and have self-confidence. In this way, women will not be victims of violence, child marriages will end, and they will gain the freedom to have children or not.

Female* voice 2

I live in a city full of fear where I question my body, my character, my identity, what I wear, my head and my hair, and I am uncomfortable with my body while walking on the street. I miss, I miss, I miss my freedom. As a woman in Istanbul, in Turkey, I miss what it means to be free as a woman in Turkey.

Female* voice 3

I want to put out many things. I don't know where to start. I want to reveal my physical expression without apologising, without fear, and especially the slightly more cabaret and more exaggerated aspects of mine. I started to eliminate them a lot as I got older. I want my infinite courage back. I would like to be able to reveal my statement. my voice, my expression. I hide a lot most of the time. We all have a terrible defence mechanism.

Inside

Residency Process

Nirah: When I first contacted Mariana, Elif and Laura, through the online residency, I couldn't have imagined the potential of this meeting. Despite the ocean of distance, the connection we were able to achieve made the whole process easier. Mariana and Laura often translated my ideas, as I couldn't always express myself fully in English. It was my first experience of international cooperation, as well as the first time developing an artistic work together, at a distance.

During the period of development of ideas (what to do? when? where? how?), it was not difficult to reconcile interests and desires to arrive at this consensual proposal. The definition of details, such as the technologies used, was a little more challenging: it was not so easy and cheap to find the equipment initially suggested here in Brazil. I had to make some adaptations, mainly because the use of masks, due to the Covid19 pandemic, was still mandatory in open spaces.

Elif: Through the online residency, I had the privilege of meeting two amazing artists and human beings, Mariana and Nirah, by whom I have been substantially nourished throughout our process and friendship. We have come to realise that we are all fond of working with the voice and body. Especially thanks to Mariana who has been integrating the notion of voice in her artistic practices and engaging with it, she has broadened our perception in many aspects by means of her deep knowledge. We have embraced the reality of our distance apart as an advantage and started brainstorming regarding how we could coalesce these notions and ideas to be melted in a pot. We were all willing to perform with our bodies in a public space and were captivated by the idea of being sounding and listening bridges with our own bodily presence of our own culture, of languages and environments that we are surrounded by. Being interfaces that connect different places, voices, languages, being three female*s in three different cities. Those were the initial points that led to this current form of this project. Concurrently, we had this urge to be provocative, to be present, to be exposed; maybe we were deprived of exposing and expressing ourselves freely and this project was the perfect moment to realize it. From my point of view, this project arose from the necessity of wanting to be heard, present, or perhaps to resist not only for ourselves but also for those who had been going through the same.

I was located in Berlin during the development of our project, and we were able to meet Mariana in person to discuss and to try out technical aspects of the project. We soldered our speakers together, sometimes she would call Nirah to update about our progress and they would talk in Portuguese for hours. It was a beautiful language to listen to even though I didn't understand a word of it. We wrapped up the project plan with Mari while eating home-made mushroom kibbeh at her place. Nirah would show her lovely cat on the webcam, sometimes she would pop up on Zoom with a winter hat on her head on relatively cold days in Brazil. All of these are worth mentioning, I guess, this warmth and connection between us.

Mariana: Our wish was precisely to stay within this place of translation from private to public, where the personal is political. It should be mentioned that we struggled a little bit with the English translation of what ended up being "to put out". In Portuguese, *botar para fora* means to put out something that's been kept. It can mean throwing up, venting, exploding, almost as an exorcism of things that were retained. In Turkish – said Elif – istifra (kusmak, vomiting), dışarı atmak (put out), patlamak (exploding). In German, Laura helped us to translate it to rausschmeißen, los schiessen (in the sense of speaking out something that is being kept), los werden (throw out). In English this connotation is not so evident, but talking to native English-speaking friends, they told us it could translate what we wanted to communicate. >What do you want to put out?< could also mean what do you want people to see as if there was something underneath it, behind the shell, hidden things that want to be shown. Or even how much of their body women allow people to have access to. In every case, there is a clear connection between what lies inside and what is made public, outwards. Moreover, they were interested in using the voice as a source material too, and I think we generated a field of resonance, sharing in distance both disparate and common experiences around those topics. Initially, we wanted to make a telematic performance, going simultaneously to public spaces with a microphone and sounding in real time the different places or the voices of people that would interact with us. They would sound in each other's mouths, a sort of broken telephone where Istanbul sounds in Berlin, which sounds in Florianópolis. We considered collecting people's testimonies in real-time, making sorts of conversations of people between the places. At this stage, we received a lot of feedback at a workshop that Georg Klein taught at Sound Studies and Sonic Arts at UdK in Berlin, where I study. Georg is one of the founders of the *Dystopie Festival*. Coincidentally I was taking his class when we were in residency, and the topic was Sound Art in public space. He and my colleagues gave very important comments that helped the process of our project during that week. Later, we decided to make an open call and use pre-recorded sounds. We could then compare how those same voices sounded in different places and also generate a more safe, controlled and intimate space for participation.

Nirah: When we defined our methodology, I bought a small amplifier of a size that would fit in my mouth and a cable to connect it to the portable battery, so that I would have the autonomy to walk around the city centre, reproducing the voices collected through a public call for women (cis, trans and non-binary), anywhere on the planet, asking them to send us audio answering the question/provocation: "what do you want to put out?"

Artistic Approaches

Mariana: I'm interested in voices. Since 2019 I've been eating voices, asking myself which are the voices that inhabit me? How are we co-constituted, multiple, weavings of difference and convergence. I have, though, placed this matter inside the body through the act of eating. Putting in. In an ongoing series of works called *alimento de vozes*, [I eat voices] by asking people to literally sing inside of my mouth or by putting speakers with pre-recorded voices inside of it. I do so while sharing my inner listening point. Through devouring or nourishing, I become, in a process of reception, internalization, filtering, transformation – a ritual of embodiment. It's a wonder on the limits between bodies and on what constitutes voice. There, I use my body as a place.

Laura: For many years, I have been studying the perception of spoken language through a musical lens, exploring strategies to integrate the human voice into compositions for instruments. In doing so, I am interested in how different modes of listening to speech are constructed – oscillating between perceiving it as pure sound and as sound charged with meaning. When I integrate loudspeakers onto the human body, I aim to emphasise that every language both arises from and simultaneously shapes a specific cultural environment. Every act of speaking is not merely a negotiation between sender and receiver; it is also embedded in – and actively contributes to – the construction of culture.

Nirah: I have been (re)searching for my own voice since I was a child. Out of fear or shame, I hid my voice for a long time and suffered from terrible throat pain, which only stopped when I started singing. Since then, I have sought to free other voices that were repressed, contained, silenced and swallowed. In the search for these voices, I also practice listening, perception, and sensitivity. And so, I see myself as a channel of reception, expression and transformation.

Elif: In the sonic imagination, the act of listening internally and externally occurs simultaneously, as a polyphony – a synthesis of sounds imagined, expected, mimicked, remembered, and perceived. They are the instant moments of sonic hallucinations drifting in between real and imagined. Sounds vanish the moment they appear; their physical energy disperses and becomes inaudible, yet they linger and leave a trace in one's mind as a form of memory or sensation – fugitive behaviour of sounds makes it even harder to catch them in the realm of imagination. Though, the voice is always present in my mind. We coexist together; the act of listening externally finds responses and merges with the inner sounding, and the voice echoes, choruses, and contrasts with what is being listened to in the moment and perceived. My inner sonic world ventriloquistically speaks through the voice that permeates and mirrors the interior and exterior sonic realms. Could it be a fundamental core in the formation of the sounds of the mind? Voice is nowhere and everywhere, catching the sounds of the mind. I overhear voices everywhere, as in seeing faces in the configurations of clouds, rocks, and patterns. I search for voice-ness in the non-vocal sound sources; I hear them in the whistling of an airstream leaking through the pipe, sometimes in a machine humming. A moment of synthesis in between my imaginary state and perceptual attention, an auditory pareidolia. My approach to voice looks at catching the sounds of the mind. I explore voice as a potential that has a role in the formation of sonic imagination to be employed for dealing with electronic music composition and performances. How accurately could I listen to the imagined sounds or create them mentally in the process of composing? How are they materialised, and through what means? In my research study of voice stands as a malleable form that carries out different affordances. Instead of privileging the content of the voice – defining what voice is or what it symbolises and represents – I fundamentally look at various forms that voice takes, essentially entangling the vocal form that permeates internal and external sonic realms and its role in this (Soğuksu, 2024). There have been many assumptions about the understanding of the voice, especially from Euro-American perspectives, as Marlene Schäfers quotes in her book Voices that Matter from anthropologist Amanda Weidman, who assumes that modern ideology of voice relies on two implicit assumptions:

> The first is that the voice constitutes a direct and authentic expression of the interior self, assuring self-presence and truth. The second, linked assumption holds that the voice functions as a transparent channel of communication whose acoustic form is inconsequential to the content transmitted.
>
> *(Schäfers, 2023, p. 9)*

Schäfers concludes that the characteristics of this vocal form contribute to "detaching voices from the subjects pronouncing them and rendering voices socially potent" (Schäfers, 2023, p. 4). In such a way, through giving this oral practice as an example, she proposes to destabilise ideas about "the voice as a direct expression of personal will and interior". She argues that voices do not always have to represent those who enunciate them, and she refuses stable associations between voice, self, and identity, but she suggests that it can also be a "technology of affect", meaning voice can be conceived as something that can be used in order to harness specific effects. It proposes that it is not only about representation but also about the multiplicity of listening, perception, affective aspects, and experience of the vocal. Throughout my research in voice studies, I realised that the perpetual presence of my inner voice inflects and infects the way I compose, the way I listen, and the way I make meaning out of things, the world, and the internal and external. In this research, it is investigated mainly through the domains of imagination and perception, materiality, and multi-sensorial phenomena – listening, sensing, feeling, producing, imagining, and perceiving the voice.

Outside

Public Space

Nirah: Although the messages were not directed at any specific person the intention was for the voices to find a place where they could be heard. In the public space, without prior notice, without a structure for wide dissemination of the performance and the project itself, the reach obtained was small, smaller than expected. For various reasons, personal and collective, including the health emergency itself, which was still in progress, we did not have the opportunity to perform this performance in other locations. Universities in Brazil were still closed, but possibly we would have a very different reach for this project by taking this performance to the university campus or even, who knows, inside a theatre in a restricted space, where attention was not in constant dispute and division, we could've expected another result.

Mariana: During the process of conceiving *speaking thr-o-ugh*, I wanted to expand on the multiplicity of voices, outwards, since we were three, in three different cities, I wanted to create something that extrapolated the inside of the body, involving not only Nirah's and Elif's bodies, but others bodies. The other two were also keen on thinking how we could occupy public space as female* bodies, and how we could establish connections between our very diverse environments. Elif brought up many points around going to the streets – disrupting, causing, shaking. She was especially interested in the interaction we would get from people in public spaces. I felt really taken by her avidness, a strong impulse for making more engaged sound art work, given the fact of how often sound art is alienated or self-referent. It was very refreshing to work with both of them around silence, voice and gender in public space.

*Being a Female**

Mariana: As female* bodies in public space, what would we like to work with? Was there a question to be answered, something that needed to be expressed, exposed in the streets? Something there to be listened to? We suddenly realised that this was precisely what we would like to work with: the presence of women* in public space, what is kept private, what is repressed, what is never shared, publicized, listened to. The rights that are threatened or never given equality, the right to exist, to diverge, the right to occupy public space as well as the right to occupy and own our own bodies. We wanted to provoke, to make all those things audible. Protest, vents, confessions. In this direction, we wrote in the open call:

The performance seeks to make womxn's voices audible, those that have often been historically silenced and oppressed.

> Our intention is to expose in public space what is often kept inside, unspoken, repressed, un-shouted, unmentionable, indigestible. Thoughts, feelings, words, laughs or screamings, vomits or other unnamed uncanny intensities. Verbal or not, a voice carries a body and a whole world.[6]

Nirah brought a lot of sensitivity around gender to our project, we all wanted to work with how silenced women's* voices have historically been. For me it was very important to have her presence as a woman from a different generation. I saw how going to the streets made so much sense for her, mostly considering the situation of violence against women in Brazil, which we, Brazilian female*s know very well. It's a situation of both historical violence, where our rights are more and more threatened and undermined within the current brutal government. Sadly, the statistics from 2021(Fórum Brasileiro de Segurança Pública, 2022) show that every ten minutes a woman was raped, and every seven hours one was a victim of femicide. I was happy that the same action was also done in my country and in Istanbul: being a woman outside of the global North is a totally different thing. My perception is that here in Europe there is still a lot to be done, but it's not so drastic. Many times, I feel powerless being here, instead of there in Brazil, unable to change things. I was very glad that this performance happened in Florianópolis through Nirah.

Through

Being Interfaces

Mariana: I walked the streets carrying voices that were not mine. As a foreigner, a stranger, a player device, a vehicle, I walked around Kreuzberg carrying vents, shouts, grunts, laughs, melodies, secrets, watchwords, songs. I was opening my mouth to convey intensities to be expressed. I had to open it widely, in order to fit the back of the speaker inside of it. Tense jaw, some cables hanging and a heavy 12V battery in my pocket pulling my coat, on an already cold autumn Saturday afternoon I was stepping into one of the most multicultural neighbourhoods in Berlin, a place of artists and immigrants. I am one of them, and, in fact, that's where I live.

I walked through my building, and my usual walk around Kottbusser Tor was different that day. I carried both my placeness and placelessness with me, while being a place for other bodies. That literally sounded through me. Re-placing myself? I walked the streets carrying voices that were not mine. Were they not mine? Who were they from? Who do they belong to? Do they belong to me, if I don't even fully belong to the place I now live in? Or do I belong then, when I carry them with me? Or, instead, how does it feel to be along those voices? How would they be listened to in Berlin, Istanbul and Florianópolis? Which are the voices we expect to hear in those spaces, and which ones are not expected? Which are the ones that try to impose themselves, or those which are successively silenced? Which ones belong to the soundscapes they sound in? Which ones are we not but one we wish they could be part of it? Which ones do we wish were louder? Which ones are the ones we want to be together with?

Being part of those voices, I question and embody the body as a place. My body was a place, not only an interface conveying intensities, but an assemblage of something else in the streets. I'm still writing and trying to put into words how I felt that afternoon while putting sentences and other vocal sounds out of my speaker-mouth. Disturbing and ignored. But also turbulent, wishful, virulent, militant, eager, vocal. Very vocal without vibrating my vocal cords. I was vibrating.

Nirah: After several days of cold and rain, finally a sunny morning arrived on the island of magic [Ilha da Magia] – the affectionate nickname of the city of Florianópolis. The wind was strong and is registered on the tape recorder I carried with me during the performance in the downtown area of the city. My son and niece acted as my assistants, filming and photographing while I reproduced voices that weren't mine, of people I didn't even know, through the amplifier adapted to my mask, instead of my mouth.

I put myself out as an instrument, as a channel, as a totem, carrying those messages with me. It could be myself talking about the fear of going out alone, the fear of going out at night, the fear of being abused. It could be my voice screaming about the tiredness of being a woman, about the weight we carry for the simple fact of existing and being who we are.

As I walked down the street, among the people, literally being a spokesperson, I felt the invisibility on my skin once again: it was as if I wasn't there, despite the voices emitted loud and clear by the amplifier. I was ignored. Even if they didn't silence me, they didn't hear me, they didn't want to hear me, they didn't care to know what I had to say or what was being said through me.

Elif: I came back to Istanbul for the public intervention after a 4-month stay in Berlin and I did my performance at Istiklal Caddesi, one day after my arrival. I was there with a small speaker inside of my mouth pointing outwards, from which

I was sounding with all the other voices which encompass the surrounding environment. Those coming out from my mouth diffusing into the most sonically diverse and multicultural environment of Istanbul. Re-bodying and surrogating collected voices of womxn, and standing for anonymous shouts, laughter, songs, murmurings, angers, vulnerabilities, intimacies, longings, calls for freedom, variety of registers, internal articulations. Each has already detached and reached me from different continents to manifest themselves and to sound thr-o-ugh me.

Mariana: Choosing to work with the mouth we could make this translation from inside to outside. The mouth is an in-between space of exchange and communication. As Brandon Labelle states in his book the Lexicon of the Mouth, in a chapter called "burp, choke, cough, gag, spit, vomit":

> There is something trying to get out; it impels the body, throttles it with hidden energy and released in choking, gagging, and coughing: an agitated breath forced out of the lungs, scraped across the larynx, to push forth over the tongue, bringing all the wet air of those deeper regions back up.
> *(Labelle, 2014, p. 44)*

I felt we. The power of what voices can carry. The plurality of voices, and further: voice as collective. I felt how the voice always searches for a listener. More than who is this? who's voicing – but also, who is this who's listening? I – or should I say, we? – were sounding with and listening to those voices. My situated position informs a listening to what comes out of my mouth, and this co-constitution in which I take part does it as well. Then it was a matter of asking how the place listened to us. Who/where is this that is listening?

Elif: I performed in Istiklal Street, one of the most important public spaces in Istanbul, due to its socio-cultural and political significance. This space has witnessed many breaking points throughout history. Hence it has become Istanbul's prime location for social activities and is popular for all sorts of parades, gatherings and protests. For this reason, this was the place where I imagined adding more layers to its current reputation. Istiklal Caddesi has always been chaotic. I was part of that cumulative sonic environment through the juxtaposition of sounds that I carried. It was an ordinary and yet vivid day in Istiklal Caddesi; I was surrounded by police with guns walking around us as usual, scattered security checks were taking place, military vehicles were parked and were ready for a possible attack, undercover police were sneaking around us. While I was walking in the street, amongst the crowds and mess, there was a moment in which I came to realise how much I have got used to living with this feeling of surveillance and control, hence oppressed and silenced. I was walking along with mixed and intense feelings, as I stepped into the crowd and amid the police, the feeling of hesitation and fear started to dominate me more and more, compared to at the beginning. Those intensities were wrapped all over my body, which surrendered to the tension. Wasn't I then the one who was oppressed under these circumstances and surroundings, while trying to free up those voices? There was this disgraceful dilemma that I was confronted with. After a while, a possibility of getting into trouble with the police forces led me to choose audio files to be broadcast in the rest of the intervention. I therefore had to eliminate some voices that had a clear semantic meaning and played non-verbal expressions instead, in case I was detained for questioning. I cannot express how guilty and painful I felt at that moment when I silenced them. I was being a hypocrite, and nothing else. I had to confront myself and the responsibility I had towards the voices of womxn I have never met. Voices were heavy and were forcing themselves outwards, but this time I had the urge of swallowing them all, deep into my body, rather putting them out as it should have done. I was full of apprehension; there was a strong tidal wave between me and the voices that I carried whilst walking down there. As I was afraid, I had the instinct of turning the volume down. Although I wanted to shout all the voices, one part of me was recessive. Wasn't I the one that was suffocated by my own saliva?

Mariana: At Kottbusser Tor, I was loud, loud as that place is, full of cars, people passing, shops, drug addicts, homeless people, old Turkish ladies with their carts full of groceries, young hipster couples, Berlin freaks, tourists, sporadic protests, police investigations, mafia, weddings, the anarchist traces at the Doppelgängeradmiral, the ever-present smell of falafel, the sound of a train passing... Kotti is mess, a mass, a mix. Maybe it's the closest place to São Paulo I could find in Berlin.

However, I was one more component of the Berlinese cacophony, adding strong testimonies in the shape of more decibels, testimonies coming mostly from Brazil, Turkey and Germany. In as many languages as I normally hear in my neighbourhood. They would put out the unwanted; narratives of sexual abuse, scorn for conservative society, patriarchy and against the Brazilian president Bolsonaro. But also groans of pleasure, uncontrolled frantic laughs, shouts of ultimate love, daydreaming, wishes. They anonymously shared the undigested, unmentioned, repressed, kept inside, unspoken, unshouted.

For me, one thing that was maybe one of the strongest aspects was kind of the responsibility of translating something very intimate into the public sphere in the sense that we got so many intimate and very strong testimonies and voices of people.

Laura: Can you give an example?

Mariana: Like someone that was saying how she has been abused when she was a child by someone in her family and how this violence is still around and it's still something that people don't talk about in a very lucid and strong way. Yeah, a very heavy story and many others like this, less individual and personal stories and more political. But I felt that those claims that the voices had to be heard were a big thing to deal with and to bring that into the public space, it was kind of a really difficult task to do. I think we did the first trial but for me it was one of the strongest aspects that we proposed this "translation", to really publicise those things, to make those things be heard. I think for me it was even stronger than dealing with the lack of reaction of people or with other aspects. But yeah, for me I think it was what we did with all those voices that we received because it was really strong.

Elif: I totally relate to that. On the other hand, I had a similar difficulty when I was doing the intervention as I had to choose which sound file I should play, considering the possibility of being questioned by the police. That was an important decision because if there had been a verbal meaning, then the police would have questioned me and seized all the equipment. The moment I realised the fact that I was actually there to shout these voices but at the same time I had to turn their volumes down or mute them.

Laura: Because on one hand, you have forces of public control, and on the other, you have a society where you can protest freely but nobody pays attention. So, it's like being ignored. And the effect is the same in the end. So, it´s silencing through silencing or silencing through not listening to, the idea of responsibility is always strong, I think, when you take other people's voices or other people's testimonies. It's very delicate anyway. But this story (the performance), it's also actually so simple, it's such a simple action that becomes so strong when you put it in the whole context and what's being transmitted. It's all amazing, I'm still amazed that this kind of action, harmless as it is, can move so much.

Mariana: Like a weapon!

Elif: Yeah. For example, this month is Pride Month. All the Pride gatherings, activities have been banned by the government in Turkey. It's just a harmless thing where people just want to gather and celebrate the freedom to be themselves, but they are all just banned. It's the same thing, actually. It was banned, but they still gathered. In the end, Turkish authorities responded with a series of attacks against activists, journalists and lawyers; it is reported that the government detained over 530 people in just 37 days and that many of those detained at Istanbul Pride reported being held for long periods in hot vehicles without water or food.

Outside the Interfaces

Nirah: The island where I live is not exactly a big city, but it is the capital of the state: it has the role of a metropolis, but it has the profile of a province. This characterisation is important, as it says a lot about how people received and perceived the intervention. I performed in the city centre, on a Saturday morning, in October 2021.

When the sounds were screams or laughter, I received a few curious stares and several disapproving looks. The children seemed more interested in understanding what was going on and came closer, being soon pulled by their mums away from me. I didn't play offensive or prejudiced messages; I didn't play messages that could be censored.

People passing by seemed to have no time to waste. Some, even wearing headphones, did not even realise what was happening around them. Even with the volume turned up, I'm not heard, I'm not seen, I'm not noticed, as if I'm not there. I do not exist in front of passersby, just as the owners of the voices I carry with me do not exist. Invisibility is painful, and that's why we persist: being seen and heard is the very possibility of existence.

If these voices do not find resonance in the public space, in the same space where merchants and musicians compete for the attention of passers-by, where can they be heard? I stopped for a while in front of a large department store, on one of the main downtown shopping streets far enough away from the competition for sonic space. A couple walk by as the voice on the recorder whispers something in a seductive tone - a whisper coming out of my mouth. The man looks at me in amazement, as if I'm speaking directly to him.

> A striking aspect of this experience is that the unfriendly reception from the public was not exclusive to the Florianópolis: in Berlin and Istanbul it was not much different. People don't seem to want to be a part of it, they don't seem to want to be aware of it, as if ignoring other realities and truths were enough to make them disappear: if I don't see the problem, it doesn't exist.

In this sense, we perceive the relevance of artistic projects that take the public space with these themes: it is fundamental and urgent to bring these issues to the attention of people. It is necessary to expose the wounds, it is vital to speak louder and louder that we need to change, that our passivity only supports the maintenance of the problems. We can no longer ignore the voices that have continually been silenced.

Mariana: Regarding the reception of the performance, despite all the power of carrying a choir, I felt impotent. The audible voices blended into the plural soundscape. Words would not stop people in the train station, they would not listen. The hysterical laughs and shouts made some people uncomfortable, puzzled, but they actually grabbed attention enough to be ignored. People would not stop to listen (only a homeless man did, he stayed very close, looked me in the eyes and joined me in the laughs). The voices actually grabbed attention enough to be ignored, precisely because I was – should I say we? – one more component of the melting pot, I hope. It fitted too much, just another performance amidst the characters and life performances happening there normally, in a place in the North where I – as someone from São Paulo – never had to walk faster, or even prevent myself from walking home at night, afraid of, for instance, having one more thing to put out.

Elif: The video documentation contains selected excerpts of the public interventions that happened in Florianopolis, Istanbul and Berlin, in October of 2021. It opens with the scene of Mariana who starts playing the audio-file from the open call "what do you want to put out?", whilst putting the small speaker into her mouth. She starts walking with slightly hesitant, but quick steps. The footage is sometimes blurry and yet sentences that are displayed are rather sharp. The perspective changes with the camera movements, from the eyes of her to the eyes of the disembodied voices. I kind of liked to see these oppositions and dualities at the beginning of the documentation: inner and outer, hesitancy and quickness, blurriness and sharpness, the mouth and the ear. Me, you and we. As three female*s in two continents and three cities shared the same feelings and carried the same migrated voices. To refer to these ideas, in the video documentation there are some juxtapositions of sound recordings of different footage in one scene and also some quick and smooth scene changes between us and the perspectives. The successive passages of each other follow chronological time, hence it inherently builds its own tension in parallel to the states of our performances. It starts with long footage sequences and starts to get more tense and drastic. Towards the end, as we perform more daringly, the tempo of the sequences gets more accelerated and sharper. Throughout the video documentation there is an interplay between perspectives, and this invites the audience to be part of the performance. It points out the notion of togetherness and unity, whereas reactions of people to our performances also play an important role in the documentation. The camera acts as the eyes of us, the anonymous voices, the people around and the audiences.

Notes

1 "mouth-street/entrance". Anzaldua often plays with the written text, writing bilingually or making up words. "Su cuerpo" means here "her" body, not "your body".
2 Documentation can be found here https://www.dystopie-festival.net/2021_B/portugues-speaking-thr-ou-gh/
3 Documentation can be found here https://www.dystopie-festival.net/2020/symposium/?lang=en
4 Documentation can be found here https://www.dystopie-festival.net/2020/showcase-online/?lang=en
5 The work can be heard under https://www.dystopie-festival.net/2021_B/memories-of-zarah/
6 From Somarumor Open Call 2021
Here, we use female meaning cis, trans, and non-binary identifying people. We are aware of the multiple forms of what being a woman is, and especially aware that the authors themselves represent a privileged group of white cis women. We are also aware that the contributions we have received are circumscribed in our networks and, in the same way, do not represent the entirety of our societies.

Bibliography

Ahmed, S. 2010. Orientations matter. In: D. Coole and S. Frost, eds. *New Materialisms: Ontology, Agency, and Politics*. Durham, NC: Duke University Press, pp. 234–257.

Anzaldúa, Gloria. 2005. La conciencia de la mestiza/Rumo a uma nova consciência. *Estudos Feministas, Florianópolis*, 13 (3): 704–719.

Bonafé, Valéria, and Campesato, Lílian. 2019. A conversa como método para a emergência da escuta de si. [The conversation as a method for the sel-listening emergence] *Debates - UNIRIO* 22: 28–52.

Carson, Anne. 1995. *Glass, Irony and God*. New Directions Publishing. New York.

Cavarero, Adriana. 2011. *Vozes plurais: filosofia da expressão vocal.* 2nd ed. Belo Horizonte: UFMG.
Eidsheim, Nina Sun. 2019. *The Race of Sound - Listening, Timbre, and Vocality in African American Music.* Durham & London: Duke University Press.
Fend, Sandra. 2020 Deutsche Frauen im Vale Europeu von der Kolonisierung bis zur Industrialisierung, Master Thesis, University Press, Vienna. Page 43.
Fórum Brasileiro de Segurança Pública. 2022. Feminicídios caem, mas outras formas de violência contra meninas e mulheres crescem em 2021. Available at: https://forumseguranca.org.br/wp-content/uploads/2022/07/10-anuario-2022-feminicidios-caem-mas-outras-formas-de-violencia-contra-meninas-e-mulheres-crescem-em-2021.pdf
Haraway. 1985. A Manifesto for Cyborgs: Science, Technology and Socialist Feminism in the 1980s. *Socialist Review*, 80, 65 cited in Pohl, Rebecca, An Analysis of Donna Haraway's A Cyborg Manifesto, Macat Library, London, 2019, p. 38.
Ihde, Don. 2007. *Listening and Voice: Phenomenologies of Sound.* State University of New York.
Labelle, Brandon. 2014. *Lexicon of the Mouth: Poetics and Politics of Voice and the Oral Imaginary.* 1st ed. London: Bloomsbury Academic.
Schäfers, Marlene. 2023. *Voices That Matter: Kurdish Women at the Limits of Representation in Contemporary Turkey.* University of Chicago Press, Chicago.
Seyferth, Giralda. 2020. *Deutsche Einwanderung nach Brasilien (Porto Alegre 2010), quoted in Fend, Sandra, Deutsche Frauen im Vale Europeu von der Kolonisierung bis zur Industrialisierung*, Master Thesis, University Press, Vienna.
Soğuksu, Elif Gülin. 2024. *catching the Sounds of the Mind: Voice and Sonic Imagination.* The Institute of Sonology.

6

HEARING THE DIFFERENCE: TRADITION, FUTURITY, AND THE BLACK TIME OF SOUNDING IN CHANNEL ONE SOUND SYSTEM'S SESSION

Brian D'Aquino

Introduction

A cold Sunday afternoon in Shoreditch. After two decades of gentrification, the neighbourhood has become a hub for East London's nightlife. In between Old Street and Commercial Street lies an endless string of pubs, bars, and music venues. Patrons crowd on the pavement as they queue, some having a smoke, all closely monitored by the bouncers. The atmosphere in the side streets is much quieter. A large van is parked outside a former warehouse, and a number of people are unloading heavy equipment from the back. At first glance, it looks like a sound rental company at work, but the equipment is far from standard.

While part of the crew waits to finish unloading, you follow others as they roll a dozen huge boxes into the venue. In one corner of the room, several loudspeakers are piled up to form a towering column. The cabinets are oversized, the design is outdated, and the surfaces are marked with scratches and dents. In the opposite corner, a second stack of speakers is now being assembled. By the wall, a man fiddles with cables and machines, assembling a strange control station. The presence of a turntable suggests a DJ booth, but the rest of the equipment is unfamiliar. The man's beard is white, and he wears a tam to cover his dreadlocks. The turntable sits atop a huge wooden box, adorned with a crumpled poster. As he continues to wire the machines, the portrait of the former Emperor of Ethiopia on a white horseback pictured on the poster is framed by unrolling electricity and signal cables. Somehow, the juxtaposition conveys a dissonant feeling.

As the control tower continues to take shape, the machines multiply, their smooth surfaces punctuated by nameless knobs and buttons. There are no

DOI: 10.4324/9781003194453-7

LED lights, no brand names. Is this audio archaeology or technology from a different future? Carefully laid cables crawl along the walls, running from sockets to machines to speakers, eventually encircling the whole room. The space is populated by middle-aged Black Rastas and younger white individuals. Everyone seems to know exactly what to do, and their relaxed confidence suggests that this scene has been repeating itself countless times before.

Shoreditch and its crowded streets now feel like a distant memory. Like vertigo, a sense of disorientation prompts us to seek confirmation about the space-time coordinates in which the scene is situated. We're in London, the current epicentre of the global reggae sound system culture and a main node in the planetary network of Black sonic practices. But what time are we in? "Time to finish stringing up the sound" – the man replies – "We're late!"

Time, Space, and Positioning

This scene described above is unfolding at Village Underground, an established music venue in East London, where the Channel One sound system holds a monthly Sunday Session.[1] The session has been regularly taking place for about a decade and is scheduled to happen again in two weeks from the time of writing. Channel One's Sunday Sessions have become a staple for the London scene, as well as for the international reggae and dub community, with YouTube footage of the dance going viral, and people flying in from all over Europe.[2] Yet what takes place at this particular London venue represents only a small fraction of the countless iterations of this scene that have occurred across hundreds of venues throughout the United Kingdom and beyond over the past 45 years. Such is the span of Channel One's service to sound system culture, a small part of which I've been fortunate enough to experience first-hand.

Channel One was established in 1979, when brothers Mikey Dread and Trevor "Jah T" decided to follow in their father's musical footsteps. Four decades later, it stands as one of the world's most well-known and respected roots and culture sound systems. Besides playing their own rig across the UK and occasionally in Europe, Mikey Dread is constantly touring to DJ in clubs and festivals all over the planet. A new generation of sound system practitioners, from Latin America to Australia, has been inspired by his music, attitude, and sonic signature. Channel One is therefore a major force in the UK sound system scene: a unique site of Black technological experimentation and sonic resistance which has not only contributed to keep reggae music alive, but also shaped the British music landscape itself, eventually projecting its influence on the rest of the globe.

I have been quite a regular attendee of the Sunday sessions in the past – under different, and sometimes overlapping lenses: as a music lover, a sound system practitioner, and a researcher. Around 2016, while completing my

PhD in the freezing London winter, far from my loved ones, something prompted me to step forward and offer to help with the sound. As a sound system owner myself, I knew that an extra pair of hands is always a plus. Still, I was surprised by how easy it was to be welcomed. "Sure", Mikey said, "see you next month, 5 pm at the gate. But don't be late".

As my time in London became more intermittent, I began planning work trips to coincide with the day of the session and sometimes travelled to their gigs across the United Kingdom. The corner of Leamington Road Villas and Westbourne Park Road – where Channel One has played consecutively for over 40 years – became my go-to spot at Notting Hill Carnival. Over the years, I've also had the honour to share the bill with them at various festivals across Europe, and to provide Mikey with dubplate mixes of my own productions (Figure 6.1).

The initial reason for me to approach Channel One was not related to an existing research interest – though, in hindsight, that may have been quietly present in the background. It also wasn't simply that I liked what I saw and heard. Don't get me wrong: as a seasoned reggae fan and sound system owner, I loved the distinctive blend of old school reggae tunes and futuristic steppers rhythms that Mikey used to play, as I valued the lyrics and the message that the MCs would carry. But it was something deeper – or, more

FIGURE 6.1 *Channel One's Control Tower* (front) (D'Aquino, 2014).

visceral – to draw me in. First and foremost, it was the unique sonic experience that they provided that led me to devote myself to Channel One more than to any other, equally inspiring, London sound system. Before any conscious decision, it was my body that responded, deeply attuned to the particular texture of sound Channel One introduced me to. Difficult as it is to articulate, this might be described as a unique combination of warmness, brilliance, and depth, consistently resonating across the multiple gigs I attended. It was this embodied response that prompted me to try and get closer. A powerful and somehow mysterious bodily affection, sound can forge unexpected proximities, bonds, and ties – in one word, communities of resonant bodies. In the freezing London winter, Channel One's sonic texture provided me with a home away from home: oddly enough, one I didn't know I was missing.

The academic interest thus ran parallel to a different kind of research: a conversation between peers (with the obvious differences in experience, knowledge, and status), largely conducted without words, and grounded entirely in practices, techniques, and technologies. The very nature of this conversation saw me in the position of a mentee, a student or apprentice – not the standard positioning for researchers, perhaps – and eventually had a significant impact on my own practice, in ways that only gradually revealed themselves. Likewise, the fieldwork this writing is based upon was made of a few formal interviews, impromptu chats and phone calls, some heavy lifting, lots of dancing, and endless hours of aural contemplation. Overall, this could be described as "observed participation" rather than the other way around, in that it was my presence that was under scrutiny.

The very nature of this ongoing conversation raised questions about how to make sense of this texture of sound, which I found myself irresistibly attracted to. Among these: how is this generated, and how is it passed on? When and where does it emerge, and towards what times and places does it move? Ultimately, how do we position this sonic texture within the broader context of a modernity that simultaneously celebrates Black sonic achievements and attempts to silence them? These questions arise from my positioning as a white researcher and practitioner immersed in a Black cultural space that is currently reshaping itself as a global culture. They reveal that positioning itself is a game of reflections – or in this case, of acoustic refractions. What we hear outside of ourselves speaks to who we are and how we position ourselves in the world.

To answer these questions, this chapter turns to the lived experience of the Channel One sound system as one example of Black sonic formations responsible for crafting an alternative sonic standard. Based on the crew members' own understanding of their practice, Channel One's sonic texture emerges as the result of a "hearing tradition" (in Mikey's own words) grounded in the embodied experience of the Jamaican diaspora: born out of necessity, passed on through apprenticeship and learning from doing. Ultimately, I suggest that Channel One's longstanding practice unveils alternative ways of being in the world that are specific to Black expressive cultures. Rooted primarily in the sonic, these ways of being are located simultaneously both inside and outside of Western modernity – a cultural formation that appropriates Black sonic innovation while seeking to relegate it to the margins.

Born Inna It

Seven decades have passed since London-based Jamaican immigrant Vincent Forbes, alias Duke Vin, launched his sound system in 1955 as the first Jamaican-styled sound system to operate in the United Kingdom. Throughout the 1960s, the popularity of sound systems as the primary source of entertainment for Black Caribbeans in the United Kingdom continued to grow. By the 1970s, virtually every Black neighbourhood would be home to one or more sound systems.[3] The most established ones would compete to perform in the few clubs catering for Black audiences and allowing them to bring in their equipment, such as the Bamboo Club in Bristol or the Four Aces in London. The rest would make a virtue of venues such as community centres, schools, church halls, and even private homes. These informal spaces are where the *blues dances* would take place.[4] And this is where youngsters Mikey and Trevor received their early sound education.

In 1979, the two brothers took over the equipment owned by their father, who used to operate a small sound system in Jamaica. Under the name of Admiral Bailey, which he continued after moving to London in the 1950s. Embracing the music and the sound system was therefore a natural way to keep a family tradition alive, slowly taking it to a higher level. As Trevor says, "We feel that we picked up, we picked the heritage from my father" (D'Aquino, 2022). As one of the hundreds of amateur, small-sized sound systems of that era, Admiral Bailey was mainly playing house parties that kids were not allowed to attend. Obviously, this enhanced the fascination that these gatherings exerted on the younger ones, as Mikey recalls.

> We was upstairs and wasn't allowed to go downstairs. The furthest we used to go was on the top of the stairs and seeing all these big people coming back and forth. Enjoying themself and partying, sound system is playing, people eating curry goat, rice, everything....Long Life beer, your Cherry B...all these things was important at that time.
>
> *(D'Aquino, 2022)*

Significantly, technology holds an important place in both brothers' memories. When asked about the *blues dances* or *house parties*, Trevor remembers exactly how the sound system would fit, and how sound would take over, the domestic space.

> House party…we're talking just one speaker, one 18-inch speaker in one corner of the room, with midrange and tops. Maybe if it is the house where you have two rooms, in the next room you will have another speaker, bass, midrange and tops in the next room…the whole house would be vibrating.
>
> *(D'Aquino, 2022)*

For Mikey, memories of the equipment predate those of the house parties, turning sound system technology into a family presence he began to become accustomed to, but was not yet allowed to interfere with, at a very young age.

> We've always had amplifiers and turntables and records from since I've known, from [being a] a youngster. It's part of the family basically, that's how we grow.…Them times there, we used to call it the front room, and that's a place where you as a kid don't go. It's a forbidden area because so much equipment is in there. So, you don't want to step on valves, wires or cables or anything like that.
>
> *(D'Aquino, 2022)*[5]

Mikey explains how the environment in which he was raised provided a unique form of education, describing his involvement in sound systems as the natural outcome of sustained exposure to a distinctive sonic atmosphere. A cultural bridge composed of sound and technology linked his two homes: Jamaica and the United Kingdom.

> For me it's not something that come up in the last ten, twenty years. I've been grown up in sound systems. We're born in sound system culture from day one, from Jamaica till now. And you have to be born inna it, to hear what the difference really is.
>
> *(D'Aquino, 2014)*

This organic and continuous exposure reflects a specific learning process, responsible for a type of knowledge that is inscribed in and transmitted through bodies and artefacts rather than stored in books or expressed through verbal communication. This is what Michael Polanyi famously describes as *tacit knowledge*, claiming that "we can know more than we can tell" (Polanyi, 1966). The idea of tacit knowledge and the process of its formation, *tacit knowing*, calls into question the privilege of the rational Western mind in making sense of or *reading* the world, and the very idea that all knowledge

FIGURE 6.2 *Setting up at Village Underground*, London 2016.

can be verbally articulated. This is reflected in the way Mikey Dread claims that "difference" must be "heard" to be appreciated (Figure 6.2).

The knowledge that Mikey and Trevor share as inscribed in their ears and bodies is not limited to technical know-how. In fact, Trevor candidly admits: "With the stringing up, as far as I'm concerned, don't put me on that side because Mikey knows it all, not me" (D'Aquino, 2022). Most of what the brothers share does not relate to sound as a physical phenomenon governed by hard science laws, nor to the technicalities of its phonographic reproduction. They rather share a knowledge of sound as material and affective matter whose perception is mediated by their physical as well as cultural personas. Indeed, Mikey and Trevor partake in a "culture", in the way the term refers to both a way of life and a range of cultural practices that build not only a shared understanding of the world but also provide a set of tools for it to be expressed. This is primarily a culture of the sonic, in the way that Michael James suggests the sonic to be "culturised sound matter" (James, 2020, p. 6). The affiliation frames the way sound is heard, felt, and made sense of, as well as what it is meant to do, and how.

One example of this tacit agreement is their shared understanding about how the system should sound. While this is challenging to express in words, it remains unmistakably clear to them, as the very foundation of Channel One's 45 years of activity. As Trevor notes, "Me and Mikey, we know how we want our sound to sound" (D'Aquino, 2022). He also adds: "You got to have quality in the way you play music". Over the years, I've heard this "quality" being alternatively described as "warm", "analogue", "clear", "authentic", or just plain "nice". Sometimes it has been expressed through what it is not, such as "not too loud" or "not ear-piercing". Interestingly, Mikey tends to describe it more via its perceived effects rather than by offering a description. "When Channel One plays, it's more quiet, but your trouser legs shake. That's how I like to play sound system" (D'Aquino, 2022). This shared understanding of sound is also the result of having been raised in an era when sound systems would play a wider social function than just entertaining youngsters on a night out, as he points out.

> Don't forget, we're coming from the era where people used to dance together. Man and woman used to dance and that's the era we're coming from. If you're coming into a dance and you're coming with your man or you come with a woman or whatever, you don't want to get your ears blown up when you're trying *fi rub*. That's very important, you get teached that way.
>
> *(D'Aquino, 2022)*

The sedimentation of such an understanding of the sonic is an ongoing process with no definitive endpoint. Mikey reinforces this when he politely refuses to be labelled "a master of his art", as I once subjected: "I don't know about being a master of the art, you just do things from day to day, month to month, year to year. It's like an apprenticeship" (D'Aquino, 2022). Despite having fully attained the status of sound system elders, Channel One's journey through sound has not yet concluded – and, in all likelihood, never will.

Hear the Difference

The staging of a reggae sound system session is visually and sonically different from that of a DJ or a live band. As opposed to a standard gig, where the audience faces the speakers while the artist plays through monitors on a stage, sound systems are usually positioned on the floor towards the performers. There is no monitoring apart from the sound system, and the crew shares the same sonic space as the attendees. At Village Underground, Channel One performs on one side of the room, surrounded by the crowd, with two speaker stacks placed at the opposite corners, permeating the room with sound. This is not sounding out but literally *sounding in* – into the room as well as into the crowd's individual and collective bodies. In terms of the performative economy of the event, the staging establishes a "tighter loop of affective modulation" (D'Aquino & Pârvan, 2021, p. 34) between performers and crowd, which amplifies the energy of the session even further. In addition, the performance of roots and culture sound systems like Channel One is not centred around mixing or *juggling* – the manipulation of the music's tempo to produce a continuous flow of rhythm – but rather the sounding of the music itself. The hundreds of people attending Channel One's sessions are certainly attracted by the quality of the music played, the MC's ability to entertain the audience (also known as "building the crowd"), as well as the possibility of listening to exclusive tunes. But first and foremost, it is the distinctive texture of sound that Channel One offers that people value most.

As selector, Mikey Dread chooses the records and plays them off a single turntable. As an operator, he shapes the sound coming out of the speakers in all its dimensions through continuous yet almost imperceptible modulation of frequencies, tones, volumes, and intensities. For Mikey, sound is far more than an object of knowledge: it is a lived experience – a distance to travel between the first and last tune he plays, a space to be built and to which to return, session after session, year after year. Although confident in his own expertise, he approaches it with the necessary respect due to a force that significantly affects the crowd and, in return, his own business. In this sense, he aims to provide an experience that is as "pleasant" as possible, and consistent with his own personal and professional trajectory. To achieve that, Channel One sticks to a completely analogue setup, including vinyl records and dubplates, hand-made speakers, and a preamp. This is not an entirely deliberate choice but rather a way to match a specific listening habit, as Mikey explains.

> The reason why Channel One plays vinyl is because of the sound it gives. The signal from the analog record, the vinyl, that warm sound. It's like when you're coming from valve [amplifiers], when you used to play valve, that's the sound you're used to. That's why this digital thing is just not for me. It's too straight in your ears.
>
> *(D'Aquino, 2022)*

The tawny speaker boxes are another distinctive feature of Channel One's set. Pictured on flyers and posters, caught on camera or video, these have become truly iconic and are associated with Channel One as much as their logo or Mikey's own face. The stack includes five different types of boxes to match the five-way frequency split. The bassline is made of so-called "scoop bins", an adaptation of a 1970s cabinet design originally aimed at bass players, still favoured by reggae sound systems around the world. When asked in more detail about how the set was conceived and built, Trevor prefers to maintain some secrecy. Nevertheless, what transpires is that the design process followed a rather unconventional pattern, wherein personal relationships, trust, and the will to experiment took precedence over datasheets and technical specifications. Eventually, the outcome was surprising even for the builder himself:

> This particular guy, he came to a Channel One session a long, long time ago. I could call a name, but the world is too big out there. This guy said, "My friend, he built a little speaker for his bedroom. If you talk to him, who knows, he might build it on a bigger level"....We went and listened to it and said, "Oh, my goodness. How would it sound on a bigger level?" We said, "You know what, build up some speakers. We're going to sort it out. Let's take a chance".... He built two stacks for us. Up to this day, we still have the same stacks.... I spoke to him not too long ago, and Mikey spoke to him. He said, "Did I build that, and after all these years, they're still kicking like that?"
>
> *(D'Aquino, 2022)*

Channel One's speakers were handmade, partly designed by twisting existing cabinet plans and partly by experimenting and testing by means of trial and error. They are certainly unique in the sense that they are one-off, but also in the way they embody a sonic expertise that does not rely on market standards but is culturally situated and collectively shared.

The most notable piece of technology in the frugal phonographic chain that makes up Channel One's sound system is a knobbed box they simply refer to as "the preamp". Unique to the reggae sound system scene, this piece of kit merges several devices, such as a phono preamplifier, a mixer, a crossover, and an equalizer, all in one stand-alone machine that enables control of the entire sound system. Preamps are not available on the standard retail market but are instead handcrafted by a few respected builders active in the scene, often customised to a soundman's own specifications. The younger crew member Nimai describes the preamp as "a custom-built box of magic". It might be added that this is not merely magic, but specifically Black magic: the product of a situated knowledge of the sonic deployed along the route of the Jamaican sonic diaspora. Besides offering a practical and cost-effective solution for sound systems, the preamp reproduces sound with a distinctive frequency response. Although perfected over time and interpreted according to the personal taste of the different builders, this "warmth" or "depth" or "weight" of sound is unerringly described as "authentic" by aficionados and professionals (Figure 6.3).

"Do remember, preamps is the brain of the sound system. If the brain is not working properly, then everything else is not gonna work properly"(D'Aquino, 2014).

Mikey's consideration unveils the sophisticated human-technological continuum that informs practitioners' own understanding of the sound system. Through a minimal manipulation of the preamp, he can shape in real-time the volume, frequencies, and tones – in a word, the very texture of sound – while at the same time controlling "ten, twenty, or thirty speakers". But, as he remarks, "[the preamp] has to work properly. And you have to work it

FIGURE 6.3 *Channel One's control tower with noise boxes (left) and preamp (right).*

nice" (D'Aquino, 2014). Complemented by Mikey's proficiency, the preamp emerges as a crucial component in the formula that allowed Channel One to ascend the ranks of the world's top sounds. When asked about the value he associates with this type of equipment, he is unequivocal, describing a preamp as a necessary feature of a reggae sound system. This assessment is not concerned with the status or exclusivity associated with the device. Rather, his judgement is once again rooted primarily in the sonic realm.

> If you want to hear the true sound of a reggae sound system, not a PA system, you can only get that sound from a preamp. That's how you work, that's how you hear things, and that's how you hear the difference. A mixer can give you a sound too clean, but to get a preamp you want that analogue sound, that warm sound that can make you say yes, I can go to a dance to hear a sound system with that warm sound playing all night.
>
> *(D'Aquino, 2014)*[6]

Together with that of the vinyl records and the handcrafted speakers, and activated by his own expertise, the sonic character of the preamp is what gives Channel One's set the "true sound" of a reggae sound system. In Mikey's own hierarchy of values, the quest for this specific texture of sound overtakes that for exclusivity, even though the latter is highly regarded in the sound system business. The preamp signature sound, alternately described as "true", "warm", or "analogue", constitutes a value in its own right. Mikey thus appears to operate according to a set of standards markedly different from those of the commercial audio industry, where authenticity comes first, and simplicity is prized.

> You only need one preamp to play a tune. Because if you have a good preamp, that is the brilliance of the sound system. How many things do you need to play one tune? That is the brilliance of the sound system.
>
> *(D'Aquino, 2014)*

The days of custom-built equipment and vinyl have long gone in Jamaica, giving way to a wave of imported US equipment since at least the 1990s.[7] Nonetheless, it is precisely through the use of the preamp and similar equipment that Channel One marks its belonging to the space of the Jamaican sonic diaspora, thus reaffirming the practice of the sound system as a Black art of sound at the apex of its global reach.

> You know that sound systems is from Jamaica, and then you have people coming from Jamaica to the UK, bringing their own tradition to here. And that's how sound systems using a preamp started. It wasn't mixer at them times or anything like that, it was just people using one preamp…so the traditional preamp has always been there, from Jamaica till now.
>
> *(D'Aquino, 2014)*

As Inge Baxmann notes, tacit knowledge is made of "sensory, emotional and perceptive experiences" that are "stored in movements, gestures and rhythms" and is eventually made material "in artifacts or nonverbal forms of communication" (Baxmann, 2009, p. 127). What Mikey and Trevor share is passed on to younger members of the crew, and it is held in high regard by Channel One's followers, which is therefore not an object – although ephemeral as a sonic texture can be – but rather an embodied, culturally situated knowledge of sound. In the case of Channel One, this knowledge encompasses practices of operating, crafting, and even marketing sound, both at a community level and in the broader context of the global music industry. Consolidated, internalised, and transmitted through apprenticeship, such knowledge is primarily embodied and only subsequently finds material expression in the technological domain – the site where it is most effectively stored, activated, and made operative. Far from being merely a technological shortcut, the handcrafted circuitry of the reggae preamp embodies a meandering aesthetic journey: the sedimentation of a distinctive sonic culture, materialised in the form of transistors.

Come to Me and Learn

Apprenticeship is the glue that keeps Channel One's co-founders, Mikey and Trevor, and the rest of the crew together. Listening to how this relationship is articulated by both sides reveals a bi-folded, dynamic, and continuously evolving correspondence that cuts across generations, ethnicity, and culture. This process is not only essential for the transmission of knowledge held by Mikey and Trevor. Sound systems are a collective effort, and having a committed and reliable crew is a key factor for the very existence of Channel One, as Trevor warm-heartedly acknowledges:

> In all honesty, every sound is different. With us, we try to make everybody comfortable around it because you can't disrespect guys who have put themselves out to come and lift speakers for you. You have to make sure that these guys are appreciated for what they do because running a sound is no joke.
>
> *(D'Aquino, 2022)*

The crew that regularly appears at Village Underground or follows Channel One across the United Kingdom can be made up of up to ten people. This includes the *box boys* who physically help with lifting the speakers and setting up the sound, as well as stallholders, truck drivers, and official or impromptu MCs. There is no high degree of specialisation, with heavy physical tasks mostly shared amongst all members. At the same time, there might be some role progression, with more experienced individuals being able to help with more technical operations. In terms of ethnicity, age, and background, the crew is rather heterogeneous, including Black and white people, Rastas and non-Rastas, elders in their 50s and younger folks in their 20s. The gender diversity is somewhat more limited, with Sista Linda, who is in charge of the merchandising, often being the sole female presence during the set-up.[8] Overall, the composition of Channel One's crew reflects the multicultural make up of their London audience and, more broadly, the positioning of sound systems as a specific, yet still male-dominated, segment of the reggae industry.[9]

In the United Kingdom and Europe, and increasingly across the world, roots and dub music have slowly ousted dancehall as the genre of Jamaican popular music most played by sound systems. The Rasta-inspired militant message seems to reflect more the feeling and ethos of a younger, socially conscious crowd, directly embodying the legacy of "rebel music" icons such as Bob Marley or Peter Tosh. Spiced up with a generous dash of electronics and synth drums, buzzwords such as "peace, love and unity" have got the better of two decades of controversies surrounding dancehall music due to lyrics accused of promoting violence, sexism, and homophobia.[10] While Jamaican-made music evolves to include influences from Africa and the United States, the fascination of the younger generation for the sound system is currently fading on the island. (Henriques & D'Aquino, 2023) A new generation of practitioners and music producers don't look to Jamaica for inspiration and mentorship, but to Black British veterans such as the late Jah Shaka, Aba Shanti-I, Iration Steppas and, obviously, Channel One.

Over the years, I have seen Channel One's crew evolving and changing. Some people have left the sound, such as longstanding MC Ras Kayleb, who moved on to pursue a solo career in 2023. Younger crew members, such as Tudor Lion and Shere Khan, have continued to follow Channel One while at the same time launching their own sound systems. Nimai is currently in the process of doing the same, while Henry Skeng is building a reputation as a music producer by supplying exclusive tunes to Channel One and other premier United Kingdom sounds. Moreover, although all members know each other, they do not necessarily spend time together outside of the Channel One sessions. This is the very context that brings them together. The crew is therefore an elastic entity, which can stretch over time to accommodate the members' own personal and professional journeys. It must also be considered that being a crew member does not qualify as a job as such, even if some remuneration can be available. It is rather a labour of love, which is fuelled by passion, commitment, and willingness to learn (Figure 6.4).

The learning processes the younger crew members partake in encompass both a knowledge of the sonic and a situated understanding of specific technologies, and even the ritual gestures that accompany them. For example, younger crew member Nimai frames his own predilection for vinyl records in a way that echoes Mikey's own opinion and is rooted in the very same set of gestures:

> I'm not a DJ, by any means. I'm a soundman, and as a soundman, I put vinyl on a pedestal, as it were…that analog sound, the experience of powering up your record player, scratching the needle to check that everything's okay, that's all part of it for me.
>
> *(D'Aquino, 2022)*

As a young man of Caribbean heritage, Nimai situates reggae music and sound systems as part of his own family heritage. His late grandfather was a sound engineer known for organising popular blues dances in Malpas Road, Brockley. Nimai's father was also passionate about music. Together, they used to attend community events where sound systems would be playing, and where Nimaieventually got hooked.

FIGURE 6.4 *Nimai wiring Channel One's noise boxes* (Brian D'Aquino, 2022).

> I was around 16. I remember going to Notting Hill Carnival with my parents, and I think I ventured off with my cousins or something, heard reggae being played through the sound systems...I got home, and remember I was so pumped up from the experience and I wanted to replicate that.... I remember saying to my dad, "How do I listen to music like we just did there?" He said, "It's pretty simple. You have to build your own speakers". I knew then that I wanted to. But it wasn't until years later that I actually started doing it.
>
> *(D'Aquino, 2022)*

As is usually the case, Nimai began his involvement with Channel One as a regular attendee, eventually becoming a *box boy*. "One day, I just said, 'I'm at home reading about speakers, building speakers. All I want to do is build a sound system. Can I help in any way? Can I be involved?' They said, 'Sure'" (D'Aquino, 2022). His current role in the crew also includes taking care of the amplifiers and the effect boxes, a natural progression given his pre-existing interest in electronics. Nimai was obviously raised in a different mediascape from that of the older generation, where the internet serves as the primary source of knowledge. Yet, as he notes, learning from the web can have its own challenges, such as being overwhelmed by the amount of available information.[11] Instead, learning by doing and through the exchanges with elders provides not only a path to navigate the information overload and the often questionable nature of online content, but also privileged access to a body of knowledge that is qualitatively distinct from that provided by any formal or DIY training.

> I've always thought of it as an apprenticeship. No one's ever said that to me, but that's the way it is…yes, I can go on the internet, and I can speak to a hundred audio engineers who are on the forum…but at the end of the day, to me, what matters is the people who are actually doing it and who have an established and recorded history of doing it.
>
> *(D'Aquino, 2022)*

This is the same approach he is currently following for building his own Nimai Hi-Fi:

> I'd scoured the internet, find some speaker plans, look at them for way too long, deliberate on which one I should go for. In the end, what helped me decide was just speaking to Mikey, and he made it quite simple for me. Keep things simple, use what works. I knew, having worked with Channel One, which sound I wanted.
>
> *(D'Aquino, 2022)*

Another regular crew member of the crew, Tudor Lion, is a young white British man who continued to work with Channel One after launching his own Creation Rebel sound system in 2018. He describes his 12-year journey with Channel One as an "organic" learning process made up of "several stages" through which knowledge would be gradually "assimilated", and his role evolved over time.

> The first days, I remember getting there and all the wires were all over the floor and all the speakers were everywhere, all over the room. I had no idea what I was doing and just trying to be of some use anywhere…eventually, the more you pay attention, the more interested you are, the more you want to learn, the more that the opportunity is there to be taught. Eventually that started to come together…the little nitty gritty is what I try to do now and leave the heavy stuff for younger and stronger people than myself.
>
> *(D'Aquino, 2022)*

Tudor also stresses that there must be a willingness to learn, an ability to pick up bits and pieces as they happen, and that it remains one's duty to connect them into a bigger picture. He describes this ability as "intuition".

> Intuition is massive, intuition and a thirst to learn. If you don't have those two things, then no one's ever going to give you the answer, no one's going to turn on your sound system for you, and no one's going to fix it when it starts to go wrong at two o'clock in the morning in Leeds, with an audience looking at you. You can't know everything yourself, and to know when to ask people is the key, but if you ask everything, you learn nothing and if you don't ask anything, you also learn nothing at the same time.
>
> *(D'Aquino, 2022)*

As a sound system owner, Tudor expands the idea of knowledge that elders possess to include sound systems as a business. For example, he considers how to navigate a particularly challenging environment like London:

> You're competing with the biggest sound systems in the world, music, theatre, cinema, events, everything...yes, you can go ahead and buy a sound system, spend £20,000 and you've had no experience on the road, you've had no experience dealing with promoters, venues, and technical equipment. All these things take a long time to pick up...the more that you can spend your time watching people who have done it for 30 or 40 years...that's the best way to learn.
>
> *(D'Aquino, 2022)*

On his side, Mikey Dread stresses how a professional relationship, which he also frames as an apprenticeship, cannot arise in the absence of a good reciprocal feeling or vibe. For him, this is the ideal setting in which the transmission of knowledge can occur.

> When you have some younger set of guys come on you is because they feel they can engage with you. They like you as a person. Forget about the sound system first. If they like you as a person, then they want to learn. If they want to do it, I always use to say, "Yes, you guys, come in".
>
> *(D'Aquino, 2022)*

The passing on of knowledge is something that Mikey shows deep consideration for, as a key factor for the continuation of sound system culture. Sound systems are often passed down through generations, typically along male family lines or to younger crew members who are gradually brought into more public-facing roles, such as selectors or operators. This is not the case with Channel One, where Mikey remains firmly at the helm.Nonetheless, his commitment to the future of the culture extends far beyond the boundaries of Channel One itself.

> Channel One won't be around forever. If you can get the knowledge now from people like I and I, then it is very important...if you can put this

> knowledge into interaction, you get the young people coming around, Black youths, white youths. That's very important because it doesn't matter what color it is. If you want to come and learn, you can come to me and learn.
>
> *(D'Aquino, 2022)*

Overall, learning is framed as a bi-folded and ongoing process, in which knowledge exchange happens in both directions. As Mikey Dread puts it, "They learn off me, I learn off them". (D'Aquino, 2022) For example, this allows Channel One to remain in tune with his younger crowd: "Don't forget, me, I'm not an internet man, I'm not a social media man, I leave it to them. That is them type of thing" (D'Aquino, 2022). Besides the personal growth and the consolidation of relationships, this mutual exchange actively challenges the generational divide that is one of biggest threats to the preservation of sound system culture, in that it pushes elders out of a constantly evolving market, while leaving youngers without any mentorship.[12]

Channel One's mentorship extends far beyond the United Kingdom. One example is El Gran Latido, based in Bogotá, which Mikey Dread personally helped build. Again, Mikey politely refuses to take any credit for the spread of sound systems in Colombia, conceiving his role more as a link in a chain that preceded and will live after him.

> We start playing in Colombia, but we start playing on the PA system. This youth Ricardo…I said to him, "Listen, you really need a sound system here"….I think it took him about two years, and one night him called me and said, "Mike, yes, I'll do it. Can you help me?" I said, "Yes, man". Now, he's a kind of father for Colombia's [sound system culture] because people are going to him now and from him other sounds in Colombia are being built. I said, "That's the whole idea". He said, "No, you are the father". I said, "No, not me. I just help you get there. Don't worry about me".
>
> *(D'Aquino, 2022)*

By means of apprenticeship, extensive touring, and a consistent online presence, Channel One's longstanding activity has contributed to the global spread of reggae sound system culture. This has inspired a new generation of practitioners who have successfully borrowed, adapted, and sometimes innovated Black techno-sonic practices to meet the needs of different cultural and social environments. For example, El Gran Latido's speakers are a regular presence at marches and demonstrations in Bogotá, something that Jamaican or even old school UK sound systems won't necessarily be inclined to do. Furthermore, El Gran Latido chose to run his set through a preamp similar to Channel One's and assembled by the same builder, with Mikey facilitating the process. Since 2020, a familiar, similar sonic texture resonates through Bogotá's streets, underground venues, and festivals, carrying the legacy forward on the other side of the globe.

Sounds from the Other Side

What Mikey Dread, Trevor, and the other crew members have generously shared provides an answer to some of the questions this piece originally intended to raise about the texture of sound that is the beating heart of Channel One's practice. Born out of organic exposure as part of a family education, passed on through apprenticeship and learning from doing, made material into and activated through culturally situated technological devices, such as the vinyl records, the handcrafted speaker boxes, and the preamp, this unique texture is a tangible manifestation of a broader body of sonic knowledge that Channel One is a living repository of. This is Black knowledge – not just because it is specific to Black music, nor because it emerges from the sonic trajectory of the Jamaican diaspora. As Julian Henriques notes, Black knowledge is black primarily because "it is made so – denigrated (in the literal meaning of the word, to make black)". And it is denigrated precisely because of what it is: tacit, embodied, nomadic, and hidden (Henriques, 2021). For Channel One, this includes specific strategies to mobilise sound to match expectations that are shared among the crew and between the crew and the crowd. What remains to be answered is the question of how Channel One's unique sonic texture – the primary manifestation of such knowledge – reflects alternative ways to inhabit modernity that are specific to Black sonic cultures more in general, and where these ways of being are situated in space and time.

From W.E.B. Du Bois to Afrofuturism, popular music has long been recognised as a vital site of Black knowledge production, emerging as a powerful tool of resistance, struggle, and collective affirmation throughout the diaspora. Within this broader context, the practice of reggae sound systems like Channel One provides a vantage point to engage critically with music as sound – matter, process, and affect – rather than text – form, meaning, and representation. For soundmen and soundwomen, music certainly plays, but first and foremost, it *sounds*. At the same time, it also brings to the fore what makes recorded sound possible in the first instance, as well as its subsequent manipulation: technology. While twenty-first-century music becomes increasingly freed from the need for a physical medium, this is not (yet) the case for sound systems. For practitioners, music cannot be separated from the massive technological apparatus dedicated to its bombastic playback.

In highlighting sonic materiality and technological mediation – two often overlooked yet fundamental aspects of musical experience – the sound system situates itself within a broader landscape of highly technologised Black sonic practices, spanning from jungle to turntablism to slowed-down hip-hop.[13] What these Black expressive cultures have in common is that they have elected the sonic as their space of action, and have engaged with it by twisting, appropriating, and sometimes misusing consumer sound technologies. By doing so, they have been able to articulate, express, and consolidate alternative lifeways from those embodied by the deadly entanglement of late capitalism and white supremacy, which is usually depicted under the apparently neutral notion of "modernity". The choice of the sonic realm shouldn't be surprising at all, given how much visual and textual means have contributed to the reproduction of the modern Western society, as well as to the racism and capitalism that are found at its very core.[14]

Here, we are reminded of Fredric Jameson's famous statement that "modernity" is not a concept as much as a narrative category (Jameson, 2012). This has been used to consolidate a hegemonic view of the world centred around a constant and ultimately unsolvable tension between the West and the Rest, affirming the former as "the universal measure of mankind" (Chambers 2022). Boaventura de Sousa Santos describes this tension as an "abyssal line" that cuts across history, geography, and epistemology to relegate everything that is located "on the other side" to a state of ontological inferiority: "Across the line, there is no real knowledge; there are only beliefs, opinions, intuitive or subjective impressions, which, at most, can become the object or raw materials for scientific investigation" (de Sousa Santos, 2007, p. 46). This is where Black sonic formations and their ways of knowing are often relegated – not the same side where technology is usually placed.

Challenging this hierarchical conception of modernity, Black sonic formations unveil alternative technological genealogies that sustain the very existence of what Stuart Hall most famously described as "other forms of life, other traditions of representation" (Hall, 1993, p. 109). They reflect alternative knowledge systems and reveal the immanence of modes of existence that question the (in)visible boundaries of Western time and space. These are literally sounds "from the other side": polyrhythmic, syncopated, oversaturated, or just "too loud" – as sound systems are often described by their detractors. Mikey Dread seems to confirm his practice to belong to this alternative lineage, which is rooted primarily in the sonic when he claims that "the difference" can only be "heard" to be "known". He articulates this further by explaining how he achieves the sound he wants without using any of the digital devices commonly used by contemporary sound engineers. "You achieve the sound you want by your ears and what you're listening to. Not everybody has the same hearing traditions, so to speak, like me" (D'Aquino, 2022). But where is this "hearing tradition" situated, and which temporality does it reflect? (Figure 6.5)

FIGURE 6.5 Channel One's electrical system, 2015 (Image: Brian D'Aquino).

A Hearing Tradition

At first glance, both the "hearing tradition" Mikey Dread refers to and the sonic texture that is its direct output are firmly anchored in the past[15]. Indeed, the whole architecture of the Channel One session appears to be haunted by the spectral weight of a past that resists temporal containment. The outdated design of the old loudspeakers, the abundance of classic tunes from reggae's golden era, and the consistent use of vinyl as the preferred format; all of it contributes to an orientation that appears to look backwards. Even the preamp can be considered a techno-fossil: the crystallisation of a tradition, the past-made-machine.

We could thus be tempted to read Channel One's activity in a way consistent with the steady rise of a "retro" trend in contemporary pop culture. Ranging from Baudrillard's (1994) idea of a "loss of history" to Mark Fisher's (2014) evocative phrasing of "a slow cancellation of the future", passing through Simon Reynolds (2011) *retromania*, these critiques converge on a common idea: retro signals a deeper dissolution of the future as a space of possibility. In the context of late capitalism, where existential precarity becomes a widespread lifeway, the turn towards the already known becomes a strategy to exorcise the anxiety of a present that appears to head nowhere.

Although these considerations are widely agreeable, I believe that they do not apply to Channel One nor to Black sonic formations at large. Tropes of return that are common in Black expressive cultures reveal a collective desire to actively engage with the past rather than reproducing it slavishly, let alone finding refuge in it.[16] This is even more the case with reggae, a music and a culture deeply shaped by a dynamic tension between past and future, reappropriation and representation.[17] Instead, by claiming to share a different hearing tradition, Mikey Dread reclaims the existence of an alternative sonic standard to which he entrusts himself.

In his book *Vibes Merchant* (Hitchins, 2016), Jamaican musician and scholar Ray Hitchins explores the emergence of what he calls the distinctive "Reggae sound" as shaped by Jamaican studio engineers since the 1960s. He notes how Jamaican music, from the very beginning, sounded markedly different from the rock and pop of the same era – genres often celebrated as the pinnacle of sound recording practices within that temporal frame.

Challenging such hierarchical readings, Hitchins proposes the notion of "recording standard", which he describes as a combination of specific approaches to equipment and recording techniques, shaped by cultural factors and aimed at satisfying the taste of the music's end user. In the Jamaican case, the end user was often the sound system crowd; the primary format was the 7-inch single; and the intended destination was the dancehall session. This broader context, where sound systems stood at the very heart of the music industry, exerted a profound influence on the formation of an alternative recording standard from that of the Western music industry, which was largely centred on LPs and home Hi-Fi systems.

In other words, reggae's distinctive obsession with the sonic materiality of the music is closely tied to the nature of its specific instrument, the sound system, and to its original setting, the streets of downtown Kingston in the 1950s, where sound would require certain qualities to attract patrons and defeat competitors. Ultimately, the intimacy between the production and reproduction of music, which was unique to the Jamaican industry of that era, with the same individuals often holding both roles with different hands, led to the development of a unique sonic standard dominating all stages of music production – from recording to dissemination. Continuing to this day, this has eventually consolidated, evolved, and travelled – unexpectedly far – across space and time.

The notion of sonic standard helps to chase the overwhelming ghost of the past that haunts Channel One's practice and situate their texture of sound in the context of a unique trajectory, which originates in Jamaica in the 1960s and still endures today. This alternative sonic standard reflects a more complex and stratified space/time, which is specific to Black sonic practices as a range of ways of being in the world that are situated simultaneously "in and out". Considering the micro space and time of sounding, where the soundwave movements occur, this is, "in and out" of Mikey Dread's own body: in his ears, trained in blues dances and house parties, as well as in the analogue equipment he favours, such as the vinyl records and the preamp.

Likewise, in the planetary space and time of racial capitalism, its stratified cultural politics and the endless circulation of tangible and immaterial goods that constitutes its bloodstream, Black sonic practices are simultaneously located within and outside the boundaries of modernity as a cultural formation which tries to silence Black sonic formations while at the same time consuming, appropriating, rebranding and marketing them to feed the global music industry. This is what happened in 2014 when Channel One risked losing its iconic spot at the Notting Hill Carnival following complaints by wealthy residents who objected that the bass levels would make their pets feel uncomfortable.[18] But it is also reflected in the generalised institutional hostility towards Black music genres such as Drill in the United Kingdom or *Baile* funk in Brazil (Figure 6.6).[19]

Black sonic formations express a temporality that exists simultaneously within and against Western time, the imperial time of colonisation, and the

FIGURE 6.6 *Channel One in session with Mikey Dread at the controls* (Brian D'Aquino, 2022).

imperative time of Black life subjugation.[20] This can be alternatively visualised as the unidirectional vector of progress or the steady metrics that regulate the system of waged labour, and is reflected in Western music's emphasis on tempo, measure, and harmony. In contrast, the time of Black sonic formations cannot be visualised at all. It can only be experienced – or, in Mikey's own words, "heard": in the syncope, the groove, the blue note, or Channel One's sonic texture. Bending time (tempo) and space (measure) at once, these are collectively engineered responses to the experience of life dominated by the abstract time of capitalism – Western time – that they simultaneously exist "in and out" of. Tao Goffe (2022) describes this alternative temporality, or lived time, as Black time, which exists "on the other side, not simply outside or excluded from time, but on the B-side of time and thus history". As not-yet-standardised time, time that was stolen and therefore is incessantly stolen back, Black time provides "a philosophical critique of standardisation as an imperial impulse" (Goffe, 2022, p. 112).

Channel One resists standardisation by resting on a different "hearing tradition"; they craft an alternative sonic standard through which they make their presence heard – in a unique way. This does not reflect any "loss of the future" but rather the persistence over time of a range of material and aesthetic strategies of survival that are specific to Black expressive cultures. Significantly, in our conversations, Mikey has never used terms such as "retro" or "vintage" to describe what he's doing or the way he wants Channel One to sound. There is no nostalgia in his choice of vinyl or the analogue preamp, let alone the desire to restore the past or stage an eerie representation of it. Rather, he does what he does because it appeals to his own ears – and to those of his audience.

Channel One's "warm", "deep", and "authentic" texture of sound is not retromania nor revivalism. It is a viable strategy to escape standardisation as the continuous theft of space and time – but not escapism. It is technological maroonage, sonic fugitivity: the result of an intergenerational practice that reorganises the past into a battering ram with which to attack a future that is yet to be written. An expert soundman, Mikey Dread, is aware of the possibilities offered by the modern sound technologies, which he politely stays away from. His choice is rather based on a different standard, centred around a distinctive texture of sound that is unique to reggae sound system culture and reflects a planetary network of Black sonic and technological experimentation where simplicity is prized and "you have to work with what you got". Reverberating just below the surface of modernity, simultaneously in and out of Western space and time, this unique texture of sound is the most striking feature of Channel One. Absolutely consistent over the years, it is always the same, and always new.

It's now 1 am at Village Underground. The lights are on; the bouncers try to disperse the crowd by directing people towards the exit. The sound of the music is just an echo in your ears, rapidly fading. As you're heading towards the exit, you hear a young person approaching Mikey Dread. "The sound was incredible! Is there any way I can help?" "Sure", Mikey replies. "Come next month, 5pm at the gate. But don't be late!"

Notes

1 At the time of the final revisions of this chapter, Channel One's residency at Village Underground is ongoing, albeit with less regular frequency.
2 See, for example, https://www.youtube.com/watch?v=AvKTMv4E09Y&t=5s (accessed January 15th, 2023).
3 See, for example, War, Paul. 2018. "Sound System Culture: Place, Space and Identity in the United Kingdom, 1960-1989". *Historia Contemporanea*, 2 (57), 349–376; Jones, Simon, and Paul Pinnock. 2018. *Scientists of Sound. Portraits of a UK Reggae Sound System*. Birmingham: Bassline Books.
4 On the blues dances, see Gates, Henry Louis Jr., 1976. "Black London". *The Antioch Review*, 34 (3), 300–317; Kew, T. 2021. "Rebel Music in the Rebel City: The Performance Geography of the Nottingham 'Blues Party', 1957–1987". In: Henry, W., and Worley, M. (eds.), *Narratives from Beyond the UK Reggae Bassline*. Palgrave Studies in the History of Subcultures and Popular Music. Cham: Palgrave Macmillan.
5 British curator Michael McMillan has done considerable work on the aesthetic, affective, and political dimension of the front room in Black Caribbean households in the UK. See, for example, McMillan, Michael. 2009. *The Front Room: Migrant Aesthetics in the Front Room*. Black Dog Publishing; McMillan, Michael. 2019. "Dub in the Front Room: Migrant Aesthetics of the Sacred and the Secular". *Open Cultural Studies*, 3 (1), 184–194.
6 I should clarify here that this was Mikey Dread's view in 2014, when we spoke – for the first time – in detail about preamps during a recorded conversation. At the time, the use of digital devices was far less widespread within the sound system scene, and the results were rarely impressive. Although I cannot confirm that Mikey has maintained this position, a glance at Channel One's control tower – unchanged since then – would appear to support that impression.
7 This was also confirmed by Jamaican sound system owner Gabre Selassie in the SST documentary film *Rockers Sound Station: Tales of the Kingston Dub Club* (2022).
8 It must be noted that Channel One also has a female booking agent, Rachel, although she's usually not present at the sessions.
9 On the female experience in the British reggae sound system scene see Charles, Monique. 2021. "Feminine Frequencies: An Intergenerational Dialogue between Soundwomen". *Journal of World Popular Music*, 8 (1), 124–140.
10 Jamaican scholars have explored the complex politics of gender and violence of dancehall lyrics, often challenging its denigratory readings. See as example Cooper, Carolyn. 1995. *Noises in the Blood: Orality, Gender, and the "Vulgar" Body of Jamaican Popular Culture*. Duke University Press; Hope, Donna P. 2010. *Man Vibes: Masculinities in the Jamaican Dancehall*. Kingston: Ian Randle; Bucknor, Michael. A. 2012. "Dangerous Crossings: Caribbean Masculinities and the Politics of Challenging Gendered Borderlines". *Journal of West Indian Literature*, vii–xxx.
11 Nimai mentioned this during the practitioner's event *SSO#8: UK Sound System Reasoning Day. Continuities and Disruptions*, organised by Sound System Outernational as part of the ERC-funded research project Sonic Street Technologies

(SST) at Goldsmiths, University of London, on the 5th of November 2022. An edited recording of the discussion is available on YouTube: https://www.youtube.com/watch?v=d8hQzDlZai0 (accessed March 6th, 2024). See also the SSO website https://sites.gold.ac.uk/sound-system-outernational/ (accessed February 8th, 2023).

12 Intergenerational disruption was highlighted as a major threatening issue by some of the most respected sound system practitioners during the event *Sounds of the Future. A Jamaican Sound System Professionals Reasoning Session*, held in Kingston Jamaica in February 2022 as part of the ERC-funded research project Sonic Street Technologies (SST).

13 For a non-exhaustive list see as example classic works such as Eshun, Kodwo. 1998. *More Brilliant Than the Sun: Adventures in Sonic Fiction*. Quartet Books; Weheliye, Alexander. G. 2005. *Phonographies: Grooves in Sonic Afro-modernity*. Duke University Press; and more recent ones such as Mackintosh, Kit. 2021. *Neon Screams: How Drill, Trap and Bashment Made Music New Again*. London: Repeater Books; Boon, Marcus. 2022. *The Politics of Vibration: Music as a Cosmopolitical Practice*. Duke University Press; Campbell, Mark V. 2022. *Afrosonic Life*. Bloomsbury Publishing USA.

14 See as example Wynter, Sylvia and Katherine McKittrick. 2015. "Unparalleled Catastrophe for our Species? Or, to Give Humanness a Different Future: Conversations". In Katherine McKittrick (ed.), *Sylvia Wynter: On Being Human as Praxis*, pp. 9–89. Durham: Duke University Press. It must be noted that racism is not exclusive to the visual realm. See for example Stoever, Jennifer L. 2016. *The Sonic Color Line: Race and the Cultural Politics of Listening*. New York: NYU Press.

15 I was tempted to rephrase in my own discussion what Mikey describes as "hearing" into "listening". As readers would certainly know, the latter is often understood as an active, conscious engagement with sound, while "hearing" typically refers to the passive, automatic process of perceiving it. However, I chose to stay faithful to Mikey's original wording. Beyond the clear commitment to honour the source of knowledge, I find his choice compelling precisely because it points to a relationship with sound that is not entirely shaped by cultural formation or conscious intention.

16 See Habiba Ibrahim, and Badia Ahad. 2022. "Introduction: Black Temporality in Times of Crisis". *South Atlantic Quarterly*, 121 (1), 1–10.

17 Sarah Daynes describes this as tenson between "tradition", "revolution", and "revelation". See Daynes, Sarah. 2016. *Time and Memory in Reggae Music. The Politics of Hope*. Manchester University Press.

18 An online petition in support of Channel One's application raised more than eight thousand signatures in few weeks and helped to persuade the Westminster council to renovate their license. See https://www.change.org/p/westminster-council-keep-channel-one-sound-system-at-notting-hill-carnival (accessed February 15th, 2023).

19 See for example Fatsis, Lambros. 2019. "Policing the Beats: The Criminalisation of UK Drill and Grime Music by the London Metropolitan Police". *The Sociological Review*, 67 (6), 1300–1316; Lippman, Alexandra. 2021. "Sonic Governance: Culturalization and Criminalization of *Funk Carioca* in Rio de Janeiro". *Anthropological Quarterly*, 94, 443–472.

20 Here I am thinking along the lines of Shaun Meyers who describes "imperative time" as the dominant construction of time as a demand or pressure exerted on Black lives, and Tao Goffe's idea of "imperial time" as standardised time, epitomised by Greenwich Mean Time. See Shaun Myers. 2022. "Black Somatics: Post–Civil Rights Movements in and Out of Time". *South Atlantic Quarterly*, 121 (1), 75–90; Goffe, 2022. "Stolen Life, Stolen Time: Black Temporality, Speculation, and Racial Capitalism". *South Atlantic Quarterly*, 121 (1), 109–130.

References

Baudrillard, J. 1994. "History: A Retro Scenario". In *Simulacra and Simulation*. Ann Arbor: University of Michigan Press.

Baxmann, I. 2009. "At the Boundaries of the Archive: Movement, Rhythm, and Muscle Memory. A Report on the Tanzarchiv Leipzig". *Dance Chronicle*, 32 (1), 127–135, p. 127.

Chambers, I. 2022. "Learning from the Sea. Migration and Maritime Archives." In Fabris, Angela, Göschl, Albert and Schneider, Steffen (eds.), *Sea of Literatures: Towards a Theory of Mediterranean Literature*. De Gruyter. Pre-print available at http://www.peopleinmotion-costaction.org/2021/02/24/2702/ (accessed 13 October 2022).

D'Aquino, B. 2014. *Interviews with Channel One Crew* [audio recording]. Unpublished.

D'Aquino, B. (2022) *Interviews with Channel One Crew Members Filmed in London in November 2022 as Part of the ERC-Funded Research Project Sonic Street Technologies (SST). See project's website.* https://sonic-street-technologies.com

D'Aquino, B., and Pârvan, O. 2021. Amplifying Street Knowledge Through Practice-as-Research: Sound System Outer National #5 in Naples, Italy. *Journal of World Popular Music*, 8 (1), 16–37.

de Sousa Santos, B. 2007. "Beyond Abyssal Thinking. From Global Lines to Ecologies of Knowledges". *Review (Fernand Braudel Center)*, 45–89, 46.

Fisher, M. 2014. *Ghosts of my Life: Writings on Depression, Hauntology and Lost Futures*. John Hunt Publishing.

Goffe, T. L. 2022. "Stolen Life, Stolen Time: Black Temporality, Speculation, and Racial Capitalism". *South Atlantic Quarterly*, 1 January 2022; 121 (1), 109–130, p. 110).

Hall, S. 1993. "What Is This 'Black' in Black Popular Culture?" *Social Justice 20*, 1/2 (51–52), 104–114, p. 109.)

Henriques, J. 2021. "Black Knowledge [SST Research 3.0]." *Sonic Street Technologies Blog*. https://sonic-street-technologies.com/sst-research-3-0-black-knowledge/ (accessed December 25th, 2022].

Henriques, J., and D'Aquino, B. 2023. "Jamaican Sound Systems and Knowledge Systems: Practice-Based Research (PBR) in Popular Culture." *Performance Matters*, 9 (1), 316–335.

Hitchins, R. 2016. *Vibe Merchants: The Sound Creators of Jamaican Popular Music*. Routledge.

James, M. 2020. *Sonic Intimacy: Reggae Sound Systems, Jungle Pirate Radio and Grime YouTube Music Videos*. Bloomsbury, p. 6.

Jameson, F. 2012. *A Singular Modernity. Essay on the Ontology of the Present*. Verso, p. 40.

Polanyi, M. 1966. *The Tacit Dimension*. Doubleday, p. 4.

Reynolds, Simon. 2011. *Retromania: Pop Culture's Addiction to its Own Past*. Macmillan

7

ETHNOGRAPHIES OF LISTENING

Settler-Colonial Violence, Erasure, and Convivial Possibilities

Manal Massalha

The following text is taken from conversations between Manal Massalha and Matt Lewis. These conversations took place between October 2023 and May 2025 (Figure 7.1).

Matt: You're a photographer, but knowing your work, you don't just parachute into an environment, you spend a lot of time making friends, working with people and it feels like photography and the camera is more than just a tool of recording. Can you tell us more about that relationship with people and sites?

Manal: Yes, for me, the camera is both a tool and a vehicle. It's the "can-opener" (Pink, 2001), that allows me to get into sites and facilitates the encounter with people. For example, on the first day of fieldwork for my PhD, which was an ethnographic piece, where I had to immerse myself in Umm al-Fahem, my research site, and investigate how the built environment intersects with the social, the racial, the legal and the political, meeting local people, learning about their everyday experiences was key. Being an outsider, however, I knew no one in the town and had barely been to it before. The camera served as my entry point; on the first day, I had nothing planned. I thought I would first get a feel of it. It was morning, just before children start school. I was stuck in traffic. In narrow, jam-packed alleyways. The sight of children carrying their school bags, trying to negotiate their way through cars caught my eye. No pavements. No traffic lights. Two or more women/mothers were standing watching children pass by. I got out of the car and started taking photos. Before I even realised, I had someone tapping on my

DOI: 10.4324/9781003194453-8

FIGURE 7.1 Aqlah's Garden, Lydd (2017): *The Land of Oranges & Olives, The Colonial Presence* series, © Manal Massalha

shoulder, "please take a photo of those children and the chaos". He was the local GP. After exchanging a few words and telephone numbers, he agreed to be my interlocutor. In the meantime, the women were looking at me. They seemed curious. This was my golden opportunity. So, I went to introduce myself, the camera, and what I was doing in town. The rest is history. The camera as a public, documentary tool had its own agency. It provoked interest in people. It wasn't used for private, snappy snap, family photos.

In relation to forging relationships with people and sites, as an ethnographer and social documentary photographer, my practice is immersive and engaging. Both ethnography and photography go hand in hand. They are co-constitutive. They are about establishing rapport and trust. Treating people with respect and dignity. Showing interest and care. Without compromising research integrity. Both to interrogate and theorise their subjects of research. I'm not a street photographer. My images are not random. They're sociologically informed. They are there to interrogate, problematise, theorise, advocate, and offer possibilities. To tell a story. The story of people and their environment. The intersectionality of gender, class, race, politics and coloniality. The social and the built environment. When the Covid-19 pandemic hit, and a lockdown was declared in March 2020, the area where I live, in Hackney, East London, was transformed. This was an historical moment that deserved and needed to be documented. Space, place, nature, people and well-being were its constituent parts. There was so much to learn from it. Having a long-standing interest in public space, green places and how important they are for our personal and collective well-being, in providing democratic places that get us out of the neoliberal, atomised, and privatised order, I thought this was a chance to record in image and sound my area that couldn't be missed. Providing recorded evidence of how convivial we are, and can be, when the conditions allow, and how desperately needed public space and nature are, especially in times of collective anxieties. By conviviality (Gilroy, 2004), I mean, our ability to live and thrive together, not despite of but because of difference. The camera was again the tool and conduit that allowed me to document, establish rapport, and talk to people. It was my theorising lens too.

Matt: What role does listening play in this ethnography?

Manal: For me, listening is being attentive to the happenings around me. It is at the heart of my ethnographic practice. To immerse myself in an environment, I've got to listen with my ears and eyes. To dig where I stand (Lindqvist, 2023). Listening helps me understand. It prompts me to ask questions: what the soundscape tells us about the social, the political, the legal, the cultural, the economic and the environmental. By environmental, I refer to the built, social, and natural environment.

So, *the towpath in pandemic times* project is squarely this. It started when lockdown was declared in March 2020. My partner became unwell with Covid. We had to isolate for two weeks. Overlooking a river and marshes and living by a beautiful hilly park and a basketball court, I had never seen the area as busy. All I could hear from the window was sounds of feet slapping, runners panting, crushed stones crinkling, wheels looping, tyres rubbing, basketballs dribbling, and birds singing. These became inseparable from the daily soundscape of the area. The cacophony of sounds and languages has never been as frequent and as diverse.

Growing up in a country where racial hierarchy and racial supremacy are the norm and the law, dehumanisation and othering of my people are embedded. Having Paul Gilroy as my teacher and supervisor, documenting conviviality became a political, hopeful act. One which helps us see there are other alternatives of being. Other possibilities. One that challenges top-down binary existence and othering. One that celebrates everyday conviviality that don't make it to the headlines. The trust in our ability to co-habit and co-exist with difference without feeling threatened. Where difference is something to celebrate. Hackney for me, especially where I live, is the site of this project of hope. Therefore, when we came out of isolation, I was desperate to start documenting my area in sound and image. The outcome was phenomenal.

As a Palestinian, whose people have been dehumanised, brutalised and othered by an exclusionary-exclusive (Massalha, 2014), racist, supremacist, settler-colonial ideology and state, it was important to me to demonstrate that people from different backgrounds, ethnicities, cultures, and religions can live and thrive together. This is not to say that colonial legacies and structural inequalities and racism don't exist in Hackney. But other forms of being, and trust do exist between communities. This is my hope. Dismantling those structures of oppression is needed to create a liberated, more equal, just, and dignified society for all.

Matt: Yeah, that feels really important and, in my mind, thinking about the difference between, a classic soundscape project, which like so much of sound art follows this idea of travelling to far-off, difficult to get to spaces, listening to rainforests and ice-caps and these kinds of things. Your project is beautiful, in that you engage with a space, if I'm not wrong, which is literally on your doorstep.

Manal: Yes, the doorstep is important. As Les Back puts it, this is about turning the ordinary and familiar into the extraordinary (Back, 2007). Into subjects of research. There is so much to learn if we just develop critical attentiveness to our immediate environment. It becomes a way of life. An approach to social research and sociology that is not limited to the physical boundaries of a classroom. It's a lively craft that pushes and stretches boundaries. Bring them into the familiar and close by. The domestic environment can also be a site of research. I started by taking short videos, for example, of my domestic space, you know, like the sound of water boiling. The chopping of vegetables. The sound of music and the radio playing in the background. This is the ambience.

Matt: Are there any other examples you could give us?

Manal: Well, whether it's in London or back home, or wherever I am, sound is another marker of an environment which suggests certain things about a space. Its past, present and future. So, in relation to my project in Lydd, which was conducted in 2017, I was interested in the urban conditions and the social fabric of the mixed Palestinian and Jewish city. Then, when I started to walk through the streets, it quickly became apparent to me how sound makes up an important aspect of this fabric. The place was haunting and haunted. Derelict buildings. Empty or emptied lots. Urban neglect. Renaming streets. Frantz Fanon's dichotomy and Manicheanism of the colonial city (Fanon, 2004) was on full display. The tension between

FIGURE 7.2 Lydd, 2017, *The Colonial Presence* series, © Manal Massalha.

erasure and assertion by force and violence of new identities to the place. The colonial violence that continues to inscribe itself in space and soundscape (Figure 7.2).

I'm also interested in urban environments, as places which allow for social interactions and sociability. The "social infrastructures" as Eric Klinenberg (Klinenberg, 2018) calls them. I'm interested in urbanity, in the heterogeneity of the city and what the built environment allows for. Communities in the city. Children in the city. What brings us together. Having children in London gives me another insight into these urban spaces, intergenerational sociability and the building of communities through space and place (Jacobs, 1961). I'm interested in the way we know them, buzzing with life and the noise of traffic, the hustle and bustle of people. The different languages in cosmopolitan cities.

When I went to Lydd it was suffocating. Sleepy. And melancholic. As if nothing but everything is happening there. The only voices you hear are the occasional passing car or the invasive mynah birds, and the airplanes, because the city is adjacent to the international airport in Israel. As a visitor to the place, when you're in those Palestinian neighbourhoods, knowing what these places used to be, you can see through its architecture that it was once something. You could feel the potentiality that was prematurely amputated and ruptured.

Matt: What would the past have sounded like? You're painting this acoustic picture of a once bustling place, because the architectural remnants of that are there to suggest that it was very different once?

Manal: Some of these, imagined, sounds come from an interview I did with Abu Muneer, one of the Palestinian elders who survived the Nakba of 1948. I'll give a bit of context to the place. So, Lydd, in Arabic, Lod in Hebrew, before 1948, when Palestinians were ethnically cleansed in their masses (Pappé, 2006) and replaced by the establishment of the state of Israel, was a burgeoning city. Situated in the centre of historic Palestine, the British Mandate turned it into a transport hub before 1948. So, a main railway passed through it, connecting Palestine to Egypt and Africa in the south and Lebanon and Turkey in the north. According to some Palestinians who remained, their survival was not by chance. They were the people who worked on the railway and were best positioned to know all about it and how to operate it. Abu Munner was one of them. He was in his early 20s in 1948 and came from a family that had worked on the railway from Ottoman times. He spoke of a city buzzing with life. Having one of the best and most efficient postal services in the Middle East. Lydd had a sorting office, newspapers coming from the Arab world would reach Jerusalem, Haifa and Jaffa on the same day. The British also built an army camp in the city, modelled around the concept of the garden city. The camp is called Al Mahatta, "the station" in English.

1948 was a transformative event. From a connected, vibrant city, Lydd becomes emptied, disconnected and isolated. A massacre takes place. The majority of its Palestinian population becomes ethnically cleansed. Only fewer than 1000 Palestinians remain. This violent history is so inscribed in the city. It's inescapable.

The soundscape of Lydd is now one of poverty and neglect. Crime and violence. Erasure and replacement. Racial tension is waiting to erupt and explode. The streets have been renamed after Zionist leaders and figures. The name of the Zionist militia unit, the *Palmach*, which was the army unit that committed a massacre, is given to one of the main roads in the old town. The roundabout outside the mosque, the site of the massacre where about 200 Palestinians were gunned down, is called the *Palmach* too. The mosque is called *Dahmash*. On top of one of its main gates, an Arabic inscription reads: "The mosque was closed down following the heinous massacre committed by the occupation forces in 1948 (1368 Muslim calendar). It was partially reopened in 1996 (1417 Muslim calendar) (Figure 7.3).

Lydd, after 48, was repopulated by what the state and the dominant Ashkenazi elite see as undesired communities. First Jews

FIGURE 7.3 1948 Lydd Massacre engraved, *The Colonial Presence* series, © Manal Massalha.

from Arab countries (Mizrahiyeem). Then came the Ethiopian Jews. The Russians. And in recent decades, messianic, religious, ultra-nationalist settlers started to flood into Lydd with a very clear agenda of what they call Judaizing the city.

Matt: How do you hear the future?

Manal: So, the near future is sadly bleak. Nearly two years into the genocide in Gaza, where nearly two million are besieged, trapped, starved murdered and bombed indiscriminately and normalised dehumanisation of Palestinians, blood-thirst, advocated by Israeli politicians and media, no Palestinian is safe from the river to the sea. One of violence and racial tension. From criminalising aspects of the Muslim religious soundscape, to asserting a violent, a messianic Jewish one. The arming of Jewish settlers and the forming of what could potentially be violent militias. Especially now, where the dehumanisation of Palestinians is reaching dangerous lows. With the genocidal campaign raging against those in Gaza and Ben Gvir, Israel's Minister of Internal Security, licenses arms to civilians and distributes them like candies (Levy, 2023) as put in *Haaretz*, an Israeli daily newspaper.

When you walk through the streets of Lydd, you see people walking by. You see Palestinians. Ultra-Orthodox Jews. Ultra-nationalist Zionists. Ethiopians. If it wasn't for what they're wearing, their attire, you would not necessarily know who's who by their features. There is so much diversity. And there is so much potential for convivial, equitable living, if it weren't for Israel's very settler colonial character. Once this colonial nomos and racial system that creates an oppressed-oppressor reality, with apartheid and supremacy, is dismantled and decolonised, then the future soundscape will be one of liberation, inclusivity, equity, and justice. It will be one of plurality, conviviality and shared humanity.

Matt: Maybe this links to silence and silencing, which is a long-time philosophical concern of sound theory, but silence seems to have a particular poignancy in regard to the space that you're looking at.

Manal: Silence, or silencing in the space we're talking about, silencing is a political act. It's very much linked to what I call the exclusive-exclusionary ideology that seeks to displace and replace. To the settler colonial project that is Zionism. In Israel, however, before this genocidal campaign in Gaza, the option of mass ethnic cleansing, displacing and replacing, was not feasible. But the act of subjugation and domination of Palestinians by other means has long been a strategy. What I call in my PhD thesis (Massalha, 2014) "technologies of control". In relation to soundscape, in Lydd, we see these attempts to silence everyday practices. Lydd is a multi-religious site. Muslims, Christians and Jews live there. Historically, for Christians it's the site of Saint George. The ritual loud call for the Muslim prayer, the Azan-Adhan, has been practised since the dawn of Islam. Back in 2017, on the morning of the first day of Eid, accompanied by police officers, Mayor Yair Ravivo stormed into one of Lydd's ancient mosques demanding that the call for prayer, which is broadcast outside from loudspeakers, be turned down. This was done within the context of trying to criminalise the loud calls for prayers and limiting them to certain times. The dawn prayer, for example. Multiple attempts at bringing into law what they call the Muizin law. Interestingly, ultra-orthodox parties in the Israeli parliament objected to the bills.

Matt: So, you could argue that a religious right is also a spatial or environmental right? The two are linked. The right to listen to or broadcast the call to prayer is an environmental right as well? I guess it's about who has control of that law and who gets to exercise that law?

Manal: In an ethno-nationalist, settler colonial context that does not see room for sharing the land, where erasure of indigenous people and their culture is at stake, the control of sonic space also becomes part and parcel of it. It is another technology of control and domination. There is a strong message here of who belongs to the space, the state and the country, and who does not. Assertion of racial supremacy in space and sound.

Matt: Let's get moving on to music, so I know you've been looking in particular at hip hop, which has been a central interest in your research, could you talk more about the relationship between hip-hop and Lydd?

Manal: So, what is interesting is that against the backdrop of this kind of colonial violence and colonial nomos, which is inscribed in the built environment, in its very social and spatial fabric, and its manifestations in deliberate neglect, poverty, crime and homicide, emerges Palestinian hip hop. It is almost no coincidence. Hip-hop historically has been a product of neglected, racialised urban spaces in the US and other places. So, *DAM*, *Da Arab MCs*, formed by artist Tamer Naffar, emerged in Lydd in the late 1990s. Lydd, his hometown, is the backdrop. The context. The inspiration for much of Tamer's music and lyrics.

So, against all these attempts to erase and silence, there is a loud, defiant and nuanced voice. His accessible, creative and innovative style became very popular. His music and writing are the soundtrack of resistance and protest against political and cultural silencing. Against the Zionist exclusionary-exclusive ideology. And the violent colonial practices of displacement, erasure and replacement.

But also inward-looking as a Palestinian in Israel, his music takes on patriarchy, homicide, homophobia, and gender violence. In his art, not only does he challenge the status quo, but he also offers more inclusive and liberating possibilities which dismantle those structures of oppression. For example, his song *The Chair* (Nafar, 2020), an everyday object that is so banal, so disregarded outside of its function, in Tamer's work takes on a whole different level. Real and symbolic. The school chair and his experience of beating and bullying by teachers, being told to sit down quietly in a chair and that he was good for nothing. The mid-life crisis of turning forty and celebrating his birthday sitting on a chair. The wheelchair of his dear, much missed dad, who suddenly became disabled after a car accident in the most formative years of his life growing up and needing the support of his dad. Who lived in a small flat. Confined by deprivation and lack of infrastructure. Dependent and reliant on Tamer to take him up and down the stairs of the block of flats. The soul-destroying experience that took his dad's dignity away. The chairs outside the family home of a murder victim, and how these are arranged, signify whether the family is going to take revenge or not. The vicious cycle of violence and homicide. The chair on a public bus, where an armed Israeli soldier is sitting... etc. So, he takes on the personal and the collective. The different layers of oppression. The military, the societal, the infrastructural, the cultural.

When I interviewed Tamer, he was intrigued by the question about the soundscape of Lydd. He had never been asked this before. Or even thought about it. He had to pause and reflect. For him, the soundscape is a mixture of sounds. Sounds of crime, deprivation, neglect, poverty and crowded housing conditions. Nails being hit to erect a new plaster wall in the house to make a new room when there is no room. The rainwater dripping from a cracked ceiling into a bucket. The chit-chat outside the building and the cracking of sunflower seed shells. The loud shouts of neighbours. The lack of privacy and a place to be. But also, the mesmerising sound of Abu Ayman, the Muazen of the Big Mosque in Lydd. Tamer says he is not religious, but Abu Ayman's call to prayer was so transcendental. So supernatural. So powerful that he inspired a sense of home. Abu Ayman was blind. He also sold treats to school children outside the school gates, during their break. Unlicensed, he faced harassment by the local authority, and Tamer and other friends would help by rushing to take all the treats so that they wouldn't be confiscated and giving them back to him later. Abu Ayman was the soundtrack of Tamer's childhood.

Matt: Are there other examples of how sound relates to issues of social, racial and environmental justice?

Manal: Sound is a key signifier which tells us about core issues in relation to social, racial and environmental justice. Speaking of environmental justice, for example, whether in an urban or rural setting, the sounds of birds, the buzzing of bees and bumble bees, the humming of insects, suggest environmental balance. I see more meadows and wildlife around where I live than when I go to the countryside, where acres and acres of land are neatly arranged. An otter was spotted this year in our local community garden. Otters, which are known to be nature's engineers, were almost extinct from urban spaces, are now returning due to rewilding initiatives and concerned citizens for whom preserving nature and rewilding are a passion. When you visit our community garden, which is run by volunteers, you feel drenched by the sound of nature. It is a rewilding site with organic food growing, where members of the community meet and new friendships emerge. Where children come and work with the land. All in direct contrast to the extractive, greedy capitalist economy that not only exploits human beings but also destroys the land and nature.

Speaking of conviviality and our ability to live with and celebrate difference, in the area where I live, I hear the call for prayer mixing with the sounds of singing and prayers coming out of Jewish seminaries, as well as the chugging of the coal boat. The sculling of rowing boats. The quacking of ducks. The sound of runners, bicycles and children. And the singing of birds. You hear the DIY bike sound system playing RnB, reggae, Jungle, Hip-Hop. The kind of rich musical tapestry that Paul Gilroy (Gilroy, 1993) talks about in the Black Atlantic.

But there are also other important sounds that we cannot afford to ignore. The sounds we hear second-hand. Mediated through Gazan journalists and citizen reporting on digital-platforms. The sounds of bombs and missiles. And the persistent hum of drones. Whether we hear them through the words, images and numbers of the tens of thousands of those being killed and injured, or through footage itself. We don't have to experience it ourselves to imagine the traumatising soundscape. The soundscape is that of genocide, that is destroying, not only Gaza and its residents, but the environment, and nature too. The terrain and the subterrain. It is a complete reconfiguration of the soundscape of our humanity. Is it one of justice? Conviviality? Inclusivity? Diversity? Environmental harmony?

Matt: I think it's really interesting what you're saying in relation to listening justice, as in the history of European listening theories, we find a great deal of prioritisation of certain sounds and music over others.

Manal: Prioritisation suggests engineering. Giving voice to some and silencing or ignoring others. It's a form of design thinking that legitimises certain desired soundscapes over others. The important question is who gets to prioritise in these processes? This is not to say that damage hasn't been done to our environment and that we don't need active intervention. It's more about how we do it. Are we being inclusive? Are we giving space for other sounds to co-exist? Are we allowing citizens to be co-producers of this soundscape? Or are we silencing them? Soundscapes should be a core element in urban planning. Soundscapes are at the intersection between the built environment and the social. They are the sonic expression of open, democratic spaces. Nature and well-being. Diversity and multiculturalism. We need an approach which allows for different ecosystems to co-exist, cross-fertilise and flow into something else. Something organic without prescription.

It's about allowing for the different sounds to be, without having to compete. So, if sound and listening are part of an everyday practice, then it should challenge these culturally centric ideas about desired sounds being prioritised. I think justice then becomes intertwined with acceptance. Unlearning. Not othering. Not giving hierarchy to certain sounds which are preferred over others.

Matt: These are great examples and important challenges to an established narrative.

Manal: So sound justice is also related to language. To what extent you feel free to speak your own language without feeling threatened? This is particularly tested in moments of racial tension. It is fundamental to exercising our democratic rights, but also an indicator of the very nature and character of the political system and ideologies we live under.

In the Israeli-Palestinian context we can't speak of conviviality as long as we have a settler-colonial system, enmeshed with apartheid and the desire to displace, erase and replace.

Matt: Could you tell us a little bit more about the role that listening plays in your current project on the marches for Palestine in London?

Manal: It is about the multiplicity of listening. The ambience of the marches. The chants. The call and response. The messages. The ecstatic, transcendental experience of coming together. To stand with the oppressed and the colonised. To demand accountability, justice, freedom and dignity. It's a somatic experience, which allows for immersion, that is healing, restorative and hopeful in the darkest of times. An experience that allows for unlearning to take places. Challenges binaries. And offers liberated possibilities.

In this project, listening is also about giving voice and a platform to the people who participate in the march. And to myself, as an ethnographer/ artist, listening and responding to the urgency of the divisive genocide. How I can reappropriate everyday technologies, such as an iPhone, to interrogate, challenge, theorise, empower, and offer liberating possibilities. Using my iPhone, I asked people around me: why they are there? Finding out and listening to people is paramount, especially in these critical, defining times where not only is a genocide is taking place, but our right to protest is also under attack. Fundamental rights such as freedom of expression, are also being undermined. There is a concentrated attempt by the government, both conservative and labour, to silence and criminalise public opposition to state complicity in genocide. In October

2023, Conservative Home Secretary Suella Braverman, called them hate marches (Syal et al., 2023).

Listening in this context becomes a moral, ethical and political responsibility. Not accepting top-down narratives but listening to the people and what they have to say. Finding out about the essence of these marches. And from what I heard in response to the simple question of why you are here today was a message of hope, love, care and assertion of our shared humanity. This is what I also call decolonisation knowledge production and engaging with urgent social and political issues. This is also reappropriation of everyday technologies, such as an iPhone, to not only document, but also respond, engage, interrogate and offer hopeful possibilities.

References

Back, L. (2007) *The Art of Listening*. Oxford: Berg Publishers.

Fanon, F. (2004) *The Wretched of the Earth*. Translated by R. Philcox. New York: Grove Press. (Original work published 1961).

Gilroy, P. (1993) *The Black Atlantic: Modernity and Double Consciousness*. Cambridge, MA: Harvard University Press.

Gilroy, P. (2004) *After Empire: Melancholia or Convivial Culture*. London: Routledge

Jacobs, J. (1961) *The Death and Life of Great American Cities*. New York: Random House.

Klinenberg, E. (2018) *Palaces for the People: How Social Infrastructure Can Help Fight Inequality, Polarization, and the Decline of Civic Life*. New York: Crown Publishing Group.

Levy, G. (2023) 'Top firearms official resigns after Ben-Gvir's appointees cut corners on gun permits', *Haaretz*, 4 December. Available at: https://www.haaretz.com/israel-news/2023-12-04/ty-article/.premium/top-firearms-official-resigns-after-ben-gvirs-appointees-cut-corners-on-gun-permits/0000018c-33cb-da74-afce-b7fb542b0000 (Accessed: 6 June 2025).

Lindqvist, S. (2023) *Dig where you Stand: How to Research a Job*. Edited by A. Flinn and A. von Rosen. London: Repeater Books.

Massalha, M. (2014) *In suspension: The Denial of the Right to the City for Palestinians in Israel and its Effect on their Socio-Economic, Cultural and Political Formation – The Case of Umm al-Fahem*. PhD thesis. London School of Economics and Political Science (LSE).

Nafar, T. (2020) *Korsi [The Chair]*. [Music video] Directed by Tamer Nafar. Available at: https://www.youtube.com/watch?v=G-G4URj-JD4 (Accessed: 6 June 2025).

Pappé, I. (2006) *The Ethnic Cleansing of Palestine*. Oxford: Oneworld Publications.

Pink, S. (2001) *Doing Visual Ethnography: Images, Media and Representation in Research*. London: Sage Publications

Syal, R., Sabbagh, D., & Stacey, K. (2023) 'Suella Braverman calls pro-Palestine demos "hate marches"', *The Guardian*, 30 October. Available at: https://www.theguardian.com/politics/2023/oct/30/uk-ministers-cobra-meeting-terrorism-threat-israel-hamas-conflict-suella-braverman (Accessed: 5 June 2025).

8

NOISE (UNK)NOW(N)

Rosie Ash

Part 1 – Sing with Me

I'll teach you a song…sing with me

Women's Group, *We Walk Tall Soundscape*

I shout: "who wants to play music with me today?" before asking my regulars if they want to play the same song as last week or start a new one. We decide to carry on working on the same one and find a corner of the community room/lunchroom/kitchen where we might be able to hear each other. We sit on two of the sofas by the window, the third taken up by one man catching up on some sleep – he slept through the whole lunch service, so we probably won't wake him. I tune up the ukuleles before handing them out, and we catch up on everyone's week. Benny,[1] who's blind, sits nearest to me as I teach him through moving his fingers to the right places on the frets and the other three members of the group sit across from me. It's always noisy, but we make it work. Today, three women are having a conversation across the room in a language I can't identify, two men are conversing near us in Arabic, and the head of the kitchen is clattering through the washing up with a volunteer.

We focus on what we're trying to do and start to play. I clap the rhythm and sing as they begin to play the chords. We have a brief pause to refresh their memory on any chords they're missing, and then we just go for it. Tamara[2] picks up on the melody I'm singing and joins in first, before Benny does the same and begins to riff his own harmonies and improvise his own words to the song. Once we've picked it up, we keep it going around and around until we're bored, or until the centre has to close, an hour after our class starts. Sometimes people across the room will join in singing, or decide

DOI: 10.4324/9781003194453-9

they want to have a go at playing. Sometimes, some of the children at the centre run upstairs to see what music we're playing, grabbing the ukuleles from us, trying to copy what we're doing before becoming uninterested and running off again. Through our musical call and response, we exercise through our bodies and voices the corporeal imitation that is one of the innate fundamentals to human existence – the attempted copying that is the basis of everyone's initial learning of the world. As Leman states: "the behaviour of another person can be understood when it can be mirrored in the subject's own action-oriented ontology...Thus, the other person is seen as having intentions, beliefs, values, and meanings" (Leman, 2008, p. 103). As we mirror and harmonise with one another, we begin to identify each other's intentionality behind their sound-making, and we understand each other differently, reforming our reality through the senses. We find and strengthen "points of connection" (Kanngieser, 2016, p. 14) through forging our own musical language. In the space of the centre, we perform with each other, for each other, improvising our own constantly shifting and limitless collective identity.[3]

The activity coordinator often asks us if we want to use the office upstairs to play, where it's quiet. We don't. We want to play music here, at the heart of the centre, in the middle of the noise: "sounds upon sounds; the overheard upon the heard" (Labelle, 2018). We stay where everyone in the community can listen in, interrupt, or sing with us.

This was several months ago. Today, I am sitting in the quiet carriage on a train from London Euston to Milton Keynes, two metres socially distanced from those around me (we're at the beginning of the COVID-19 pandemic), and someone is playing their music out loud from their phone. I can hear it over my noise-cancelling headphones, and glance over quickly but say nothing. I try not to listen and focus on transcribing a sound piece that I made in collaboration with the women's group at the London-based destitution centre where I ran the above music class – a centre made for and with refugees and asylum seekers who cannot work and/or afford to pay for their basic requirements.[4] Several are homeless, and many are stuck in an enforced poverty limbo, waiting for the Home Office to grant them indefinite leave to remain so they can work, move, and live.[5] Some are stuck in destitution for years:

> You know it's not easy, I'm here in this country sixteen years, six - teen years...I should be taking my son to school, you know? But – I say things turned the way they have turned, I've tried, what can I do? I'm not gonna scream or just cry...because I'm strong...I'll just be like, I hope there will be a way for a normal life, you know? Because that would make me happy.
>
> *(Ash, 2019)*

Here, she emphasises the strength that she has had to cultivate in order to continue to fight for "a way for a normal life" where she can take her son to school. Many of the women speak about wanting their voices to be heard and listened to, encouraging reflections on the world which has denied so many people access to the lives they want to build: "what I am saying, I would like them to listen to me, and try and attribute it to their daily experience" (Ash, 2019). In this chapter, I hope to do this. I will listen to the spoken experiences of the women and reflect on some of the plural conclusions that this community might reveal about the current realities of being a migrant in the United Kingdom. Through examining how the country's services and provisions for asylum seekers have continually failed the people they are supposedly designed to help, I hope to expose inherent "aporias and contradictions" fundamental to the historically formed oppressive structures that this country sustains (Fisher, 2007, p. 27).I aim to reveal how the process of moulding persons who come to the United Kingdom into asylum seekers allows political imperatives to fictionalise the reality of people's embodied (often traumatic) experiences and the plural consequences of this. The sonic, it seems to me, is a crucial way to access this knowledge that has remained unheard:

> Sound figures as a vital articulation or lens onto the body and the tensions of its social performance, by making corporeality explicit: guttural, abrasive, intimate, explosive, vocal, and assertive, sound may amplify the inherent forces and drives of physical experience and what it means to be a body.
>
> *(Labelle, 2015, p. 104)*

Through using sound to emphasise that our corporeality is often at "tensions [with our] social performances" (Labelle, 2015, p. 104), we can identify how we all shape ourselves, and our bodies, in order to fit within the legal and social boundaries enforced upon us. Identifying these tensions can also make evident discriminatory patterns of how legal and social boundaries have instead become disabling restrictions for particular communities or groups of people. I intend to show that sound and voice are central to uncovering and resisting such oppressive structures. All the individuals recorded are happy for me to use the sound piece that I created in collaboration with them as a guiding intersection throughout this chapter, but will remain anonymous as many are still in long legal battles whilst they seek refugee status.[6]

Throughout my reflections, I will endeavour to remain self-aware of my positionality and thought processes, for whilst it is not possible "to do research that is uncontaminated by personal and political sympathies", I can explicate where my sympathies lie (Becker, 1967, p. 240). Consequently, I have chosen to use passages of self-reflexive writing throughout to make evident my embodied position of writing and explore by what means the women's stories have intersected with my own personal and political reflections. In discussing polyvocality in relation to their sound work, Evan Ifekoya states:

> I am the author of this work, and I speak from a subjective position, which is all we can do as human beings. There is no objective positionality. Everything I say and do is influenced by my surroundings, my upbringing, cultural stuff, my gender, my sexuality, my race… all of these things. I wanted to be explicit with that but also acknowledge that every singular subjective is made up of a whole myriad of subjectivities.
>
> *(Ifekoya, 2018)*

Ifekoya refers to their work as an expression of both their individual voice and an intersection of a plurality of voices and vocal influences. Whilst they can only express their own positionality, this is not to say that parts of their experience do not have complex and important relations with the experiences of others. I, too, can only speak from my subjective position. I am a resident of the United Kingdom; I was born here and have no lived experience of migration or forced displacement. It is important to acknowledge that I have only heard the stories and experiences quoted throughout in part and as a second-hand listener whilst volunteering at the destitution centre. Fortunately, I have been able to work with the women's group and other people at the centre for over a year, but I attempt to remain aware of this privileged access point throughout my reflections on my conversations with the women. I am laying no claim to their words, instead exploring the potential of listening to a plurality of voices – voices that have been politically and sometimes legally disarmed.The destitution centre is home to one of the most diverse and resilient communities that I have had the joy to be a very tiny part of, an "unlikely public", of people who:

> Hover unsteadily and ambiguously in the open, shaping themselves within quotidian spaces and locations often between communities, languages, and even nation-states, to form volatile coalitional frameworks; that draw from resources found in collective intelligence, shared skills, popular traditions, and from the energetic knowing of the senses…constituted by those that cannot wait for systems and states to catch up with them – they are, instead, gaining traction through the thickness of relations, giving expression to *life in the making*.
>
> *(Labelle, 2018)*

People come to the centre (a converted three-storey house on a residential street) to volunteer, collect food parcels, use computers, attend classes, access legal advice, make music, chat, and eat meals together. Many people at the centre "hover unsteadily" between communities and nation-states, as they wait for the Home Office to decide where they legally belong. However, it's not uncommon for individuals to return to the centre after they have been given indefinite leave to remain and have no immediate need for the centre's assistance. The *unlikely public* of the centre, however volatile and unsteady, continually and limitlessly draws from their collective experiences, shared strength, and resilience. The space of the centre is not without its imposed legal limits and restrictions: you can only come to the centre once you have begun the official process of seeking asylum, or as a registered and trained volunteer, like me. Hence, the "systems of law and control", that are continually "at pains to monitor, capture and manipulate the intrusiveness of noise" (Labelle, 2018), permit the centre to be a safe, although monitored, space for asylum seekers to find resources but restrict their activity to this crowded, underfunded, and oversubscribed centre The support is vastly inadequate for the people it supposedly serves: "migrant help, they gave us a roof over our head, and little for food, which is not enough" (Ash, 2019). However, the people in the centre do not stagnantly wait for the "systems and states to catch up with them", even if their lives have become restricted (Labelle, 2018, p. 15). Their *unlikely public* provides traction towards "a way for a normal life", a *life in the making* (Ash, 2019). This chapter is not an examination of the exact processes that asylum seekers have to pass through in order to be granted refugee status, nor a piece of written advocacy.[7] I am not providing a retelling of these people's stories, but focusing on the importance and potential that lies within the act of these people speaking out about their experiences and the urgency of them being heard. Similarly, I do not intend to conflate the diverse experiences of people in entering the United Kingdom to seek refuge, but instead reflect on the intersections of their experiences as told from their individual, embodied points of view through focusing on a sonic understanding of their stories. Their words that I will quote were shared in the context of a sound piece that was played as part of a live theatre performance at Southwark Playhouse, *We Walk Tall* (November 2019), that was devised and performed by the women with the facilitation run by *Compass Collective*.[8]

Part 2 – We Are the Same

We are the same; it doesn't matter what colour we are, what background we are, what language we are speaking. (Ash, 2019)

- Niamh,[9] Women's Group, *We Walk Tall Soundscape*

With the exponentially increasing risks of climate change disasters, and now, as I write this, in the midst of a global pandemic, we are in a worldwide moment of temporal and spatial vulnerability. The banal operation of our everyday lives seems more fragile than it has ever been before. As the "harmonious relationship between humans and nature breaks down", it appears "the challenge is to rebuild our conceptual scaffolding to reflect these novel realities" (Sayre, 2012, p. 58).

It has become evident that the current treatment of the planet cannot persist without ending many human lives. Particularly in danger are those in communities that are already disproportionately exposed to concentrated risks to their health, freedom, and life: "the Anthropocene is predicated on exploitation, colonialism, slavery, and genocide, and any claim to an equitably "responsible" humanity subjugates these structural brutalities" (Kanngieser, 2015, p. 2). This *novel* reality, or perhaps the reality that governing forces (who maintain such *structural brutalities*) are only now realising the implications of, is that the world is now on a path to being treated as a "planetary mine" (Labban, 2014, pp. 560–576).

"Geographies of extraction have become entangled in a global apparatus of production and exchange that supersedes the premises and internal dynamics of a proverbial world system of cores and peripheries defined exclusively by national borders" (Arboleda, 2020, p. 5).

Whilst it is undeniable that the global climate crisis is in some way intertwined with the exponential growth of global trade networks that were *instigated* through processes of colonialism that exploited borders between nations (such as the oil and gas industries) (Bonneuil & Fressoz, 2016), as we move further down the current climate crisis trajectory, national borders are revealed to be imaginary as all relations between the leaders of nations become essentially strategic value exchanges. The notion of the world as *planetary mine*, now "signals the emergence of a more coercive, centralized and authoritarian configuration of late neoliberal statecraft", rendering national power obsolete (Arboleda, 2020, p. 6).

The horizontal comradeship of nationalism has begun to break down, as has been revealed by the recent divisive elections in the United Kingdom and United States of America, and further evidenced by global protests such as Extinction Rebellion that have created new strains of comradeship independent of nation (Anderson, 1983, p. 7).

Despite the leave campaign for Brexit continually promoting ideas of a populist nationalism to target EU immigration as a disruptive force that "couldn't (or shouldn't) be controlled", the vote to leave the European Union again further divided the nation (Denham, 2019).Energy issues are "at the epicentre of not only the geopolitics of empire and the global climate crisis, but also the more banal everyday reproduction of capitalist social life" (Huber, 2008, p. 113). Without interrupting the hegemonies and hierarchies that are fundamental to the endless reproduction of the *banal everyday*, it becomes impossible to imagine a future that diverts us away from the ongoing worsening of the global climate crisis. Donald M. Lowe states:

> All body practices – not just production and consumption, but social reproduction, gender construction and sexuality, and even psychopathology – have become commodified to such an extent, that the satisfaction of our diverse bodily needs is reconfigured by the requirements of flexible accumulation.
>
> *(Lowe, 1995, p. 2)*

It is this ongoing flexible accumulation that drives us closer to an earth that is merely a *planetary mine* and is no longer of service to our "diverse bodily needs", or even to our survival. Even sleep, Jonathan Crary has argued, is on a path to being consumed by capitalism (Crary, 2013). We move towards "an illuminated 24/7 world without shadows…the final capitalist mirage of post-history, of an exorcism of otherness that is the motor of historical change" (Crary, 2013, p. 9). I use Jonathan Crary's notion of a 24/7 world as emblematic of a world turned *planetary mine* – a result of "the plural consequences of neoliberal globalization and of longer processes of Western Modernization" (Crary, 2013, p. 8). Crary's concept of a 24/7 world feels a deeply embodied way to discuss "some of the paradoxes of the expanding, non-stop life-world of twenty-first century capitalism", without privileging such concepts as *the West* or *late neoliberal capitalism* as having specific explanatory significance. As Patti Lather succinctly states:

> Situated in the crisis of authority that has occurred across knowledge systems, my challenge has been to make productive use of the dilemma of being left to work from traditions of research legitimacy and discourse of validity that appear no longer adequate to the task.
>
> *(Lather, 1993, p. 683)*

It is undeniable that much of the theory that relies on concepts such as *the West* illuminates revealing conclusions within an appropriate context; however, such broad and all-encompassing terms wield an implication of monolithic authoritarian power run by *the West* via *capitalism*. This is not the case. Instead, a wide-reaching and complex "matrix of domination", –a conceptual model developed by Patricia Hill Collins – governs a 24/7 world. The *matrix of domination* is built from "any and all systems of oppression that mutually constitute each other and shape people's lives" (Constanza-Chock, 2020, p. 20). Therefore, it is how our everyday lives are shaped and constituted through social relations that must be examined in order to identify the contradictions within the *matrix of domination*. By assuming our modes of existence to be in some way natural rather than reliant on historically formulated hegemonies and hierarchies that support the endless and exponentially growing flow of capital, it becomes impossible to envision any other possible "conceptual scaffolding" (Sayre, 2012, p. 63). Our personhood and all of its "body practices" have become reconfigured and distorted by capitalism and its demands, but this is not to say the capitalist structure of society is so deeply ingrained that our structures our impenetrable. Colin Dayan states that "in legal rationales, realities are created" (Dayan, 2011, p. 246). However, our reality, the *banal everyday* of an ever-more 24/7 world, is accidental and contingent. The legal rationales that govern us are neither necessarily right nor unchanging, and neither is our reality. As stated by Mary Midgeley, "the Marxist pattern of complete final revolution is not at all appropriate here. We do better to talk organically of our thought as an ecosystem trying painfully to adapt itself to changes in the world around it" (Midgeley, 2011, p. 7).

An incoming 24/7 world is a possible future that emphasises how the lived experience of our everyday lives painfully adapts to complex and contradictory capitalist structures and their required "legal rationales" in order to survive. It is within these irresolvable internal dynamics of our world ecosystem, that the detrimental effects of the *matrix of domination* can be identified; through seeing where people have painfully adapted their bodily needs in order to survive and fit within the world's systems (system's that are presumed to serve our bodily needs) we might begin to see how such oppressive structures can be subverted. Brandon Labelle talks about the complex difficulties, but also the opportunities of the disturbances of forced displacement:

> Migrancy and the potent disturbances it delivers...may enable the establishment of new social formations, especially those grounded not only in narratives of homeland, but rather according to world-making activity... such positions and productions may work to undermine or supplant the embedded violence of nationalism.
>
> *(Labelle, 2018, p. 113)*

Asylum seekers and refugees have often adapted more than most in order to leave behind nation-states within which their bodily needs were not provided for. The vast majority have gone hungry, physically and mentally exhausted themselves on journeys, and left behind families and communities in an attempt to find a place where their most basic bodily needs are sufficiently protected by law, and yet are continually failed by multiple nation-states in securing this. The daily lived experience of migrants and their unique positions are, therefore, central to understanding the most extreme failures of the world's *conceptual scaffolding*. The climate crisis has begun what is likely to be one of the largest crises of migrants in history; it is widely accepted that there will be an estimated 200 million climate migrants by 2050, "representing a ten-fold increase over today's entire documented refugee and internally displaced populations" (Brown, 2008).

Hence, it seems that the lived reality of migrants will become a much more widely shared experience as the crisis grows and ensues. Despite this, the ongoing political scapegoating of refugees and asylum seekers as the reason for various failings of our *conceptual scaffolding* has remained a theme across the world, but particularly in the United Kingdom since the 1980s (Kaye, 1998, pp. 163–182). There have been growing restrictions for refugees across Europe for several decades:

> Through the use of specific language, the status of refugees and asylum seekers has been questioned and doubt has been cast upon their "genuineness" in terms of the UNHCR definition as interpreted by the UK authorities.
>
> *(Kaye, 1998, p. 167)*

The changes in language discussing refugees have happened over time via political speeches, policymaking, and through the media. Where refugees were previously seen as needing unquestionable protection under international law, the introduction of the term "economic migrant" being used interchangeably with the term refugee (and sometimes "immigrant") by policymakers in the 1980s began to imply that the people coming to the United Kingdom might be lying about their reasons for arriving in the country. This not only cast doubt on the validity of refugee voices, but also created a huge surge in discussion around refugees, emphasising to the wider public that migration to the United Kingdom was "a problem" that required political management. In saying, "we are the same", Niamh does not mean that she assumes some automatic or pregiven sameness; we have different backgrounds and subjectivities that have informed our experiences up until now. Niamh means that through "find[ing] and strengthen[ing] points of connection between different communities and publics, [we can begin] to forge shared languages around everyday struggles and concerns" (Kanngieser, 2016, p. 14). Through hearing voices that make "corporeality explicit: guttural, abrasive, intimate, explosive, vocal, and assertive" (Labelle, 2015, p. 104). We can identify how every single one of us "hover[s] unsteadily and ambiguously in the open, shaping themselves within quotidian spaces and locations" (Labelle, 2018, p. 15). We are all subject to a reality created by legal rationales, and we all shape ourselves (to differing extents) in order to find a place within the systems of control that we can continue to exist within. (Dayan, 2011, p. 246). Through listening to the vocalised expressions of these women's experiences:

> We can begin to discern the force of an alternative, post-racial identity based…on will, inclination, mood and affinity…a chosen, recognisably political idea of kinship that is all the more valuable for its distance from

> the disabling assumptions of mechanical solidarity based on either shared blood or shared land.
>
> *(Gilroy, 2010, p. 92)*

This is not to say that such kinship can escape the *matrix of domination* but to suggest that the powerful force of a "recognisably political idea of kinship" lies in its ability to illuminate new potential futures that could, and should, serve all of us. Through forging passages of connection based "on will, inclination, mood and affinity", new coalitions that are further from "disabling assumptions of mechanical solidarity" can be formed (Gilroy, 2010, p. 92).

Through listening to the vocalised experiences of migrants that have been forcibly placed in positions between national borders, we can begin to identify how they have formed *unlikely publics* that continually transform and find expression for "life in the making" as it changes at an ever-increasing rate. (Labelle, 2018, p. 15). The strength of these communities to form and continually evolve, in spite of the legal rationales that force them into ever more unstable spaces, holds potential for real futures that can form in resistance to such legal rationales and the structural brutalities they uphold. The volatile nature of our environment has never been more evident; our fast, collective adaptation is more necessary.

Part 3 – I Wish People Can Be Heard

> I wish people can be heard. Be heard and you get ears that can listen to you, you know? And find you as a human being...What I am saying, I would like them to listen to me, and try and attribute it to their daily experience.
>
> *(Ash, 2019)*

As I sit in this quiet carriage of the train, overhearing the noise of those sitting opposite from me, I become frustrated with their interruption: "encounters with noise are transformative, sensuous and intensive: noise is often felt as well as heard, and known through feeling" (Thompson, 2017, p. 9). In disturbing me, their sound "invades my space – it disrespects borders, thereby making explicit the intensity of territory", and becomes, as I subjectively designate it, noise (Labelle, 2015, p. 24). The territory of the quiet carriage defines their act of sound-making as a questioning of or even disregard for the borders of the quiet zone and, by being loud enough to overcome my noise-cancelling headphones, the borders of my personal space – the privilege of quietness is no longer afforded to me. In displaying their disregard, they alert us to the previous silence of the carriage, illuminating "systems of law and control", that "are at pains to monitor, capture and manipulate the intrusiveness of noise", (Labelle, 2018, p. 68), or "noise by definition is that sound which occurs where it should not. It finds its way in, to disrupt the particular setting" (Labelle, 2010, p. 47). To make a sound when instructed to be silent is to bring into question the social dynamics of the space at that moment in time, and the structures that surround and arrange our interactions. Through disrupting the quiet, they cause me "to shift from a state of inactivity to alertness", my quiet space is invaded, and I am forced into an encounter with the noise (Thompson, 2017, p. 9).

I could speak up, "extending this body that acts, that expresses my subjectivity as an intensity of fragments and vectors", but in doing this, "I am also made vulnerable" (Labelle, 2018, p. 64). Through using my voice, I reveal myself to be a body:

> In the same way that voice does not merely communicate information, the voice goes beyond the linguistic meaning of the words, and in their voices, we hear their emotion, personality, accents, breath – they become an imaginable living entity.
>
> *(Gallagher, 2019c, p. 283)*

If I were to ask the people making noise to be quiet, I would simultaneously choose to be an enforcer of the systems of law and control that govern and restrict us both *and* join the noisemaker in disrupting the setting, putting myself at risk of someone perceiving me as a threat to the social order. Of course, it may "matter greatly who behaves strangely, or occupies the position of a stranger within moments of disordering interruption, for such cuts and breaks may also incite abuse or criminalization" (Labelle, 2018, p. 71). Noisemaking makes audible the multiple hierarchies of power in any space and reveals the effect of these hierarchies on our bodies. Hence, this encounter begins to make explicit the blurred and subjective boundaries between sound and noise, as well as the transformative (not necessarily positive) potential that lies within the act of making noise. By breaking from the written script of the space, the noisemakers reveal such a script to be capable of being suspended: "an audible interaction between writing (the dictates of law) and noise (the suspension of law)". In this complexly structured space, to disturb one another by breaking with this silence is to initiate the possibility of "knowing each other in the extreme", (Labelle, 2010, p. 109), because we are abruptly distanced from the "disabling assumptions of mechanical solidarity" (Gilroy, 2010, p. 92). My voice reveals me to be "an imaginable living entity", (Labelle, 2010, p. 62) and consequently, challenges the people in the space to address and respond to my subjective "intentions, beliefs, values, and meanings" (Leman, 2008, p. 103). To make a noise is to disobey the imposed silence, and bring into question the true power of the hierarchies that govern the space. A tannoy announcement might play, undoubtedly read by a "female-sounding voice: in the British context, this voice is clipped, upper-middle class, brusque, sensible" (Power, 2018). In the territory of the quiet carriage, the voice of the corporate powers (or the voice they have chosen to represent them) governs this space and its announcements are never noise, but necessary sound. An artificial voice – recorded, regulated, coercive in its polite instruction – has authority and takes precedence over each of our embodied voices, and signals "any number of structures and portals of control" (Labelle, 2018, p. 64).

The voice interrupts us all as it speaks over us, but it's inflexibility also signals the inability of the voice to access the potential of noisemaking. The voice's detachment from an imaginably *present* body confines the voice to a bodyless "then" and "there" ...the incorporeal...the immaterial" (Güçbilmez, 2007, p. 153). The scripted "aural clutter" of the perfect tannoy voice detaches the words from a body as they "float monotonously over commercial and privatised zones", (Power, 2018). Implying they can never be interrupted or interfered with, and that the zones they govern will carry on exactly as they always have, without disturbance, change, or evolution. Hence, the artificial announcements cannot access the affective capacity of noise to "modulate, perturb and transform", (Thompson, 2017, p. 6), as they instead insist everything is as it *should* be. Society and it's bodies and voices constantly change and shift, making unavoidable noise in the process, but the voices of the "systems of law and control", that are at "at pains to monitor, capture and manipulate the intrusiveness of noise", (Labelle, 2018, p. 68), confine themselves to the past, a there and then, which cannot transform as quickly and flexibly as a noisy crowd. This is not intended to equate noise with political resistance, as this would fail "to account for the predatory nature of capitalism", (Thompson, 2017, p. 124). But to highlight the productive capabilities of noise to activate improvised transformations within society through creating "sensuous and intensive" interactions that provide new ways of knowing "through feeling" (Thompson, 2017, p. 9). Noise is that which disturbs and disrupts the individual, abruptly alerting them to their position within the present collective. In doing so, noise and it's disruptions can connect a new *unlikely public*, capable of transforming the present into previously unimaginable futures.The continuation of our daily reality as a society governed by *systems of law and control* relies on the vast majority of individuals believing their perceived place as a member of society and their corresponding identity (be this national/sexual/gender identity) to be in some way natural and not entirely constructed (and therefore capable of being completely deconstructed). Through the historical written word, the effect of a reality rendered *natural* by each nation's "legal rationales" (Dayan, 2011 Dog, p. 246) is the ongoing patriarchal and racist (and otherwise discriminatory) construction of a reality which all of those living in mainstream society partake in. This complex relationship between legal rationale and reality is paradoxical and reliant on the continuation of hegemonies of the past. The making of societal laws has remained confined to that which can be communicated in written symbols and languages. This reveals the ocularcentric co-ordination of the English language and the resistance, even reluctance, of visually organised societies to attempt documentation of the other senses. I refer to visually organised societies as those that proclaim observation as central and necessary to the revealing of objective rationalism, as is the case in the United Kingdom. R. Murray Schafer states that the ear only "gave way to the eye as the most important gatherer of information" in "the days of writing", and was associated with the beginning of religious portraiture, as "it was not until the Renaissance that God became portraiture, previously he had been conceived as sound or vibration" (Schafer, 2012, p. 103).

Despite the evidently affective potential of sound – able to invoke a spiritualistic notion of a god – processes of learning-by-ear have been disregarded in favour of placing vision at the axis of knowledge and construction of history through prioritising the written and recorded "objective" word (Levin, 1993). Crary highlights the centrality of vision in "an illuminated 24/7 world without shadows…the final capitalist mirage of post-history" (Crary, 2013, p. 9). This suggests the legal rationales that define our personhood and govern our realities are thoroughly intertwined with the hegemony of vision. Dayan examines "the poor treatment, the entombing of the living…as something culturally inevitable in that the past haunts us"(Dayan, 2011, p. xvi.) Therefore, if the hegemony of vision and the priority of the written word maintain the power of the past over the present, the ocularcentric also seems central to the *matrix of domination* that continues to *entomb the living* and oppress people today.

Decolonial and Afrofuturist writers such as Paul Gilroy and Kodwo Eshun highlight that "struggles against racial hierarchy have contributed directly and consistently to challenging concepts of the human" (Gilroy, 2010, p. 59). Their work towards imagining futures beyond racist structural brutalities requires even further dismantling of the hierarchies that govern our humanity and personhood. Vision is central to Du Bois" definition of double consciousness: the "sense of always *looking* at oneself through the *eyes* of others", and is inherently attached to the white (in this case American and early twentieth century) visual construction of a "dark body", as other (Du Bois, 2015). However, until recently, the hierarchy of the senses has remained largely undiscussed in relation to much scholarly writing on racism (Gilroy, 2010, p. 57). The films of the Black Audio Film Collective, which explored the experiences of Black and Asian diasporas in the United Kingdom, had a huge focus on the importance of sound, voice, and noise's resistive potential. Hence, it seems that challenging the hierarchy of the senses, one that lauds vision as truth, might indeed challenge other notions of *natural* hierarchies and *natural* forms of societies' operational structures (Gilroy, 2010, p. 59). In order to imagine new futures, we must first strengthen points of connection through creative embodied encounters (perhaps through noise) in order forge new "shared languages" that strengthen publics and their collective voices (Kanngieser, 2016, p. 14).

Part 4 – Background Noise

If then, "the residues of terror are never really dead and gone but, through the terms of law, survive and always find new bodies to inhabit, new persons to target", (Dayan, 2011, p. xiii).

This presents the urgency of the need to identify the current targets, and question how we might halt the endless evasion of responsibility. Laws materialise the memories of past atrocities against the oppressed, allowing the *matrix of domination* to continue to hold power over new victims whilst the screams of the unheard voices – those who have suffered in these histories, the background noise – are often lost. In many ways, the corporeality of the actual human experience is left largely unrecorded. Even music, a listening activity that many of us take part in every day, is categorised in observational terms, as written sheet music with a list of permitted and labelled expressions to accompany the mathematical values of the notes. Even with the invention and wide dispersion of recording technology, sound and music are "still accessed in terms of metadata such as the name of a composer or the title of a song, but not in terms of how it sounds or how it feels" (Leman, 2008, p. 1). Jacques Attali concludes that endless attempts to make experienced events, such as listening to music, into visual objects "amounts to an attempt to make people believe in a consensual representation of the world" (Attali, 2009, p. 46). Hence, the visual scripts of music, space, language, symbols, etc., allow a passivity in viewing which may be conceived as objectivity, occurring as the result of being several steps removed from the experience the representations are trying to communicate. This evokes Guy Debord's notion that "all that once was directly lived", would become "mere representation", in such an ocularcentric society of the spectacle, disabling individuals' abilities to examine reality in any corporeally *felt* way (Debord, 1994, p. 1). Our visual, passive systems for communication of knowledge do not predicate any need for discussion and questions with other bodies about their "intentions, beliefs, values, and meanings" (Leman, 2008, p. 103). This allows a "kind of individualism that treats people, and indeed other organisms, as essentially separate, competitive entities" to remain the "dominant ideology", rejecting and "ignoring the fact that competition can't get going at all without an enormous amount of cooperation to make it possible" (Midgeley, 2011). This individualism, necessary for capitalism, is therefore presumed to be natural. Crary states that in a 24/7 world, "the fragility of human life is increasingly inadequate...sleep has no necessity of inevitability" (Crary, 2013, p. 9). As we try to sleep, we accept:

> A form of time that leads us elsewhere than to the things we own or are told we need...One hears sounds of traffic, a dog barking, the hum of a white-noise machine...also the flood of anxieties, fears, doubts, longings, imaginings of failure.
>
> *(Crary, 2013, p. 127)*

In these times of inherently vulnerable human existence, the universal experience of trying to get to sleep, we shut our eyes, and in doing so, our hearing "leads us elsewhere". We absorb the background noise around us as sound; sound that irritates, angers, startles, and soothes. We begin to listen to our surrounding sonic environment not as signals from "the things we own or are told we need" but as overheard sounds (Crary, 2013, p. 127). As these sounds reach us, they "may disrupt our sleep, causing us to shift from a state of inactivity to alertness" (Thompson, 2017, p. 9). In our moments of trying to enter into sleep "involuntarily, one reclaims a sensitivity or responsiveness to both internal and external sensations with non-metric duration" (Crary, 2013, p. 126). With our eyes closed, we cease trying "to look upon the world", and through "the flood" of our emotion, remember that "the world is full of noise and that death alone is silent" (Attali, 2009, p. 3). Through this moment of rejecting the driving forces of capitalism and instead attending to our bodily needs of sleep, we escape from a 24/7 world of "mere representation", (Debord, 1994, p. 1), to an "elsewhere", that returns us to an "energetic knowing of the senses" (Labelle, 2018, p. 15.) In the process of ceasing to listen for signals from the 24/7 world, we attend to the background noise instead. Recently, there has been an increase in academic writing around the topic of noise:

> In the discourses of media theory, there is a growing sense that noise matters to understandings of our contemporary, highly mediated society, in which noise's presence is suppressed, restrained and combated but...never eradicated.
>
> *(Thompson, 2017, p. 2)*

In many of these discussions of noise, "noise veers towards becoming everything and subsequently, nothing. Noise and sound become largely synonymous" (Thompson, 2017, p. 1). Broadly, noise is defined as unwanted sound, interference with signals, often connoted with disturbance or unpleasantness. In terms of communication transmissions, noise is to be avoided at all costs: "terms such as clean, neutral and transparent are used to describe apparatuses that (supposedly) exactly reproduce sounds without "colouring" them, that is, over- or under-emphasising certain frequencies" (Gallagher, 2019a, p. 277).

Space does not permit me to explore all the rich semantic connotations of the language used around sound and noise, but I wish to emphasise the inherent attachments of noise to that which interferes with the communication from *authorised* mediators. The classifications between sound and noise are political decisions about what information is made legible.[10] Perhaps no clearer line has been drawn between sound and noise, signal and interference, than within military settings. Theorists such as Kittler suggest that these violent military implications are an inherent part of our communication technologies. For more information on the military definitions between noise and sound, see Steve Goodman, *Sonic Warfare: Sound Affect, and the Ecology of Fear* (Cambridge: The MIT Press, 2012). As I transcribe this soundscape made with the women, I can make legible that the words they speak signal information, but in the act of reducing the "bodily fullness and immediacy of [their] speech", and the background noise created by the space of our recording to the "seamless coherence of writing", I deny that such *background noise* holds value (Gallagher, 2019b). By beginning this chapter with the audio of the soundscape, I hope to make evident the importance of this noise. The voices of the women, with their "huffing and puffing, thick accents and stumbled half-sentences...bursts of laughter, stutters, coughs and guttural mumblings", illuminate just as much, if not more, than the signals of their words about the women and their relationship to the world (Gallagher, 2019b). It is this background noise that signals their connection to their bodies and consequently transforms them into "imaginable living entities" (Gallagher, 2019a, p. 283).

Marie Thompson, through detaching noise from negative connotations of *unwanted* or *undesirable* interference, attempts to "present noise as a productive, transformative force-relation and a necessary component of material relations" (Thompson, 2017, p. 4). Noise is unavoidable, and like Labelle, Thompson takes noise as a moment of disturbance that is both necessary and holds potential to be transformative. As our journey towards sleep takes us "elsewhere", (Crary, 2013, p. 127), we reclaim a deeply felt and embodied listening state. Crary goes as far as to suggest that this moment is the ultimate space for "the imaginings of a future without capitalism", suggesting that such a future "begins as dreams of sleep" (Crary, 2013, p. 128). Through attending to the background noise with our eyes closed, we access an "energetic knowing of the senses" (Labelle, 2018, p. 15) and perhaps (however, briefly) escape the metric and monitored trappings of a 24/7 mode of existence. This is not to claim noise as the epitome of political resistance, as it has been sometimes radically claimed, but to suggest noise as both a subject and methodology, an ongoing disruption of an already multiply ruptured narrative; a productive starting point to examine the world in an embodied way that "betrays the binary; it is unfaithful to dualistic thinking, perturbing neat categorizations and distinctions" (Thompson, 2017, p. 7). In disturbing us, noise attaches the sounds we are hearing to our sensuous experiences, and it is this moment of connection between noisemaker, listener, and embodied reaction that I want to emphasise as something deeply powerful.

Part 5 – The Right to Live

Everybody got the right to live the life they deserve. I'm not talking about basics. I'm talking about the life they chose. The lives they wanted.

- Women's Group, *We Walk Tall Soundscape* (Ash, 2019).

The people dubbed "asylum seekers", represent the extreme manifestation of existing on the boundary between their biophysical embodied and felt reality and their socio-politically imposed identity. On their entrance to the United Kingdom, their voices and words become more than personal soundings. From the moment they declare their being in this country and seek asylum (or are unwillingly documented), their every word is recorded and bound to them as their "objective" identity in order to define their legal status. In many instances, the corporeality of their experience is disarmed, bodily evidence of torture even disregarded, in order to make quicker, simpler asylum decisions (Nye, 2016; Townsend, 2010). In far too many cases, this means that in becoming an *asylum seeker*, these individuals feel a complete rejection of their lived reality, even an erasure of their existence, by the nation-states and communities they are attempting to join. Their voice ceases to be their own and is taken from them in quotes, out of context, often to assist the Home Office (wherever possible) in bringing down immigration statistics to aid the political ventures of current governments. The individuals attempting to seek refuge in the United Kingdom become what Dayan might refer to as "*depersonalized.* Disfigured as persons, they are then judged outside the law's protection or most susceptible to its violence" (Dayan, 2011, p. xiv). Through the restrictions of the legal system, which are continually increased and tightened, asylum seekers and refugees are refused the right for their voice to be heard, as sound, confining any such speaking out against the failings of the United Kingdom's system as intrusive noise. Consequently, the "systems of law and control", take great "pains to monitor, capture and manipulate the intrusiveness", of the people that they have labelled asylum seekers. (Labelle, 2010, p. 68). Before these people even speak, their voices are confined to a category of noise that must be controlled and minimised, rather than heard.

Amid the pandemic, the depersonalisation of refugees and asylum seekers is more felt than ever. "I don't think we exist", was a quote from Abanda, interviewed by *The Guardian*, "a refused asylum seeker who advocates for migrant rights" (Goodfellow, 2020). Through labelling migrant's voices as noise, *systems of law and control* disarm the world-making potential of migrants" unique positions, because "such positions and productions may work to undermine or supplant the embedded violence of nationalism" (Labelle, 2017, p. 113). To acknowledge migrants, exist and matter is to begin to dismantle "a world system of cores and peripheries defined exclusively by national borders" (Arboleda, 2020, p. 5). National governments are rendered powerless in a truly globalised society in which we are all migrants. As the world becomes a *planetary mine*, the climate crisis and its resulting influxes of migrants will monumentally transform how power operates as "a more coercive, centralized and authoritarian configuration of late neoliberal statecraft", begins to emerge (Arboleda, 2020, p. 6). In confining and minimising the *background noise* of refugees, subjecting them to such depersonalisation that they "don't think [they] exist", the power embedded in their extremely flexible and adaptive unlikely publics is refused recognition:

> It is necessary to ban subversive noise because it betokens demand for cultural autonomy, support for differences or marginality: a concern for maintaining tonalism, the primacy of melody, a distrust of new languages, codes, or instruments, a refusal of the abnormal....They are direct translations of the political importance of cultural repression and noise control.
>
> *(Attali, 2009, p. 7)*

In making asylum seeking into a political issue, the United Kingdom has failed and continues to fail to uphold the basic human rights of people seeking refuge in this country, instead using the ever-growing restrictions on freedom of movement and migrant rights as a performative display of "cultural repression and noise control" for the sake of demonstrating some semblance of power over an ever more global world.

There is a blank refusal by political parties and the wider public to acknowledge the powerful potential of the voices of those who have forged their own places in the world outside of the systems of "legal rationale" and in doing so, have created extremely flexible and adaptive communities with limitless potential. Through rejecting the acknowledgement of these people as individual and conscious actors choosing, often in very difficult circumstances, to come to the United Kingdom, the depersonalising narrative inflicted on them allows not only their ongoing mistreatment but also allows underlying racist narratives to remain untouched and unquestioned. The process of depersonalising asylum seekers as they arrive in the United Kingdom begins with their individual voices being used against them. Such symbolic processes of "cultural repression and noise control" are upheld and implied to be politically necessary for the continuation of society as it *should* be – governed by systems of power that uphold the matrix of domination and consequently drive us all ever closer to a world controlled by a "coercive, centralized and authoritarian configuration of late neoliberal statecraft", a 24/7 world of mass depersonalisation. (Arboleda, 2020, p. 6). For an asylum seeker or refugee to reclaim their own voice and to demand to be heard (however quietly) is to begin to interrupt this process.

As a British Citizen, I will continue to attempt to hear and amplify the voices of those who have been denied humanity by the law or socio-political processes in the hopes that, through doing this, we can begin to "rebuild our conceptual scaffolding" in a way that *actually* serves our shared realities. "Novel realities", have always been experienced by those forging their own path to reclaiming their personhood outside of the law's protection, and diasporic communities have always formed their own conceptual scaffolding to survive this (Sayre, 2012, p. 63). To operate in a 24/7 world in which we are all on the path to being depersonalised, we must turn to those who have already experienced this process in order to learn how to survive, or ideally, divert the path to such mass depersonalisation.

Epilogue – We Got Released

I Said Yes is a poem shared with me by one of the women in the group:
I said yes
They say leave
I said stay
They are hostile
I am friendly
They say no
I said yes
They discriminate against people
I feel equal
They spread stress
I meditate and relax
They have security
I seek peace

They say no
I said yes
They have authority
I have god
They create fear
I love safety
They detain people
We got released
They refuse us
We go to court
I fight back
I will win
I said yes, yes, yes. Women's Group, *We Walk Tall*, as shared during the live show in November 2019.

As I edit the sound for this piece, I add an echo to punctuate her last line, hoping that the strength of her words will resonate with those in the

Southwark Playhouse theatre as much as they have with me. Through these words, she has highlighted the individuality of her experience and the difficulty of her personal fight with the ongoing restrictions and legal battles that she faces. The affective simplicity of her chosen words simultaneously highlights the pervading oppressive power of *them* – those with authority and power to interrupt the telling of her own story – against a singular "I". As she runs through her own positive affirmations, "I said stay...I am friendly...I said yes", *they* break up her thoughts and flow. Her voice makes apparent the endured loneliness and almost impossible amount of strength required to fight a legal battle, which is a huge part of many asylum seekers' entrances to the United Kingdom, as they fight to be able to stay in the communities and families that they have migrated here to join. It is as if her words are not being heard, as *they* blankly refuse each of her statements, not engaging in a dialogue, but instead replying to her with a list of unexplained denials. Discussing "the sight of citizens turned refugees", Dayan states, "they were alone, although they were surrounded by people. They had claims on our sympathy, but nonetheless remained powerless, prey to the vagaries of coercion, legality, and control" (Dayan, 2011, p. 210). However, whilst she begins the poem "prey to the vagaries", she does not remain powerless.She finds the strength to finish the poem in the last paragraph as she switches from "I love safety", to "We got released...We go to court". She draws strength from an unlikely public that shares her experience, or aspects of her experience, alongside her. This becomes enough to "fight back", and her poem finishes with her own words echoing back to herself, without interruption from *them*. Even though it is unclear if she has won her case by going to court, it is as if she has already won through gaining the support and shared strength of a community. Even if they continue to deny her legally granted safety and peace, she has begun to find that in the humanity around her. From finding others that have either shared parts of her experience or are willing to share a part in her experience from now on, several voices have joined her initial vocalisation, in a resonant "yes, yes, yes", declaring their own freedom. I ask one of the women, "if you were to teach me something right now, what would it be?" She smiles and says she would "maybe teach me a song". Immediately, she's on her feet, singing:

> Merrily, merrily shall I live now, un-... because I've gotten my freedom... under the blossom that hangs on the bough, merrily, merrily, shall I live-...You know freedom has no limits, freedom means you are free to do what you want to do, you understand? He's gotten his freedom so he's so happy, say merrily, merrily shall I live now....
>
> *(Ash, 2019)*

Notes

1 All names changed to preserve anonymity.
2 All names changed to preserve anonymity.
3 For more on playing music as (per)forming identity, see Nicholas Cook and Richard Pettengill, "Introduction", *Taking It to the Bridge: Music as Performance* (University of Michigan Press, 2013).
4 An asylum seeker is a person who has crossed an international border in search of safety and has applied for refugee status under the 1951 UN Convention, whereas a refugee is someone who has gained refugee status under this convention. For more information on the definitions throughout the asylum seeker process, as well as useful statistical data about refugees' experiences.
5 United Kingdom's asylum support is currently around £5.66 a day (after a £0.26 a day increase due to COVID-19), providing asylum seekers with less than £40 a week to live off – they are not legally allowed to work until granted refugee status, and so many people rely on the handouts from various centres around London.
6 As an asylum seeker, any public discussion about their journey to the United Kingdom or the difficulties they have faced risks harming their legal cases – there are various rules and regulations about claiming asylum "as soon as is reasonably practical" and within the first European country they arrive in. If they publicly expose that they may have breached one of these regulations, or that their words might no longer align with what they said in their initial "screening", this can be enough for their asylum claim to be denied, or even for them to be placed in a detention centre indefinitely. See https://www.gov.uk/claim-asylum/after-your-screening and https://righttoremain.org.uk/toolkit/detention/ for more information on the risks involved in the process of claiming asylum.
7 Great work is being done in these fields by organisations such as Free Movement, (see https://www.freemovement.org.uk/how-to-claim-asylum-in-the-uk/), Refugee Action (see https://www.refugee-action.org.uk), and Right to Remain (see https://righttoremain.org.uk).
8 See https://www.compasscollect.com for more information.
9 All names changed to preserve anonymity.
10 Perhaps no clearer line has been drawn between sound and noise, signal and interference, than within military settings. Theorists such as Kittler suggest that these violent military implications are an inherent part of our communication technologies. For more information on the military definitions between noise and sound, see Steve Goodman, *Sonic Warfare: Sound Affect, and the Ecology of Fear* (Cambridge: The MIT Press, 2012).

References

Anderson, Benedict, *Imagined Communities: Reflections on the Origin and Spread of Nationalism* (London: Verso, 1983).

Arboleda, Martín, *Planetary Mine: Territories of Extraction Under Late Capitalism* (London: Verso, 2020).

Attali, Jacques, *Noise: The Political Economy of Music*, trans. by Brian Massumi (London: University of Minnesota Press, 2009).

Becker, Howard S., 'Whose Side Are We On?' *Social Problems* 14.3 (1967), pp. 239–247.

Bonneuil, Christophe, and Jean-Baptiste Fressoz, *The Shock of the Anthropocene*, trans. by David Fernbach (London: Verso, 2016).

Brown, Oli, 'Migration and Climate Change,' *IOM Migration Research Series* 31 (2008), 1–60.

Constanza-Chock, Sasha, *Design Justice: Community-Led Practices to Build the Worlds We Need* (London: MIT Press, 2020).
Cook, Nicholas, and Richard Pettengill, 'Introduction,' in *Taking It to the Bridge: Music as Performance*, ed. by Nicholas Cook and Richard Pettengill (University of Michigan Press, 2013), pp. 1–19.
Crary, Jonathan, *24/7: Late Capitalism and the End of Sleep* (London: Verso, 2013).
Dayan, Colin, *The Law Is a White Dog: How Legal Rituals Make and Unmake Persons* (Oxford: Princeton University Press, 2011).
Debord, Guy, *The Society of the Spectacle* (New York: Zone Books, 1994).
Denham, John, 'Nationalism in England Is Not Just a Right-Wing Nostalgia Trip,' *The Guardian*, 13 August 2019. https://www.theguardian.com/commentisfree/2019/aug/13/english-nationalism-brexit-remain-and-reform
Du Bois, W. E. B., *The Souls of Black Folk* (London: Yale University Press, 2015).
Fisher, Jean, 'In Living Memory...Archive and Testimony in the Films of the Black Audio Film Collective,' in *The Ghosts of Songs: The Film Art of the Black Audio Film Collective*, ed. by Kodwo Eshun and Anjalika Sagar (Liverpool: Liverpool University Press, 2007), pp. 16–30.
Gallagher, Michael, 'Audio Recording as Performance,' in *Non-Representational Theory and the Creative Arts*, ed. by Candice P. Boyd and Christian Edwardes (Singapore: Palgrave Macmillan, 2019a)
Gallagher, Michael, 'Voice Audio Methods,' *Qualitative Research*, (2019b), pp. 449–464 doi:10.1177/1468794119867549 [Accessed 22 June 2020].
Gallagher, S. *Action and Interaction* (Oxford: Oxford University Press, 2019c) .
Gilroy, Paul, *Darker Than Blue: On the Moral Economies of Black Atlantic Culture* (London: Harvard University Press, 2010).
Goodfellow, Maya, 'The Hostile Environment Is Creating a Coronavirus Crisis for Britain's Migrants,' *The Guardian Online*, 30 March 2020. https://www.theguardian.com/commentisfree/2020/mar/30/hostile-environment-covonavirus-crisis-britain-migrants [Accessed 4 April 2020]
Güçbilmez, Beliz, 'An Uncanny Theatricality: The Representation of the Offstage,' *New Theatre Quarterly* 23.2 (2007), pp. 152–160.
Huber, Matthew T., 'Energizing Historical Materialism: Fossil Fuels, Space and the Capitalist Mode of Production,' *Geoforum* 40.1 (2008), pp. 105–115.
Ifekoya, Evan, 'Polyvocality: Evan Ifekoya,' *Mousse Magazine* (2018), http://moussemagazine.it/evan-ifekoya-gasworks-2018/
Kanngieser, Anja, 'Geopolitics and the Anthropocene: Five Propositions for Sound,' *Geo Humanities* 1.1 (2015), 80–85.
Kanngieser, Anja, *Experimental Politics and the Making of Worlds* (London: Routledge, 2016).
Kaye, Ronald, 'Redefining the Refugee: The UK Media Portrayal of Asylum Seekers,' *The New Migration in Europe*, ed. by Khalid Koser and Helma Lutz (London: Palgrave Macmillan, 1998), pp. 163–182.
Labban, Mazen, 'Deterritorializing Extraction: Bioaccumulation and the Planetary Mine,' *Annals of the Association of American Geographers* 104.3 (2014), pp. 560–576. https://doi.org/10.1080/00045608.2014.892360 [Accessed 3 April 2020]
Labelle, Brandon, *Acoustic Territories: Sound Culture and Everyday Life* (London: Continuum, 2010).
Labelle, Brandon, *Background Noise: Perspectives on Sound Art* (London: Bloomsbury Academic, 2015).
Labelle, Brandon, *Sonic Agency: Sound and Emergent Forms of Resistance* (London: Goldsmiths Press, 2018).

Lather, Patti, 'Fertile Obsession: Validity After Poststructuralism,' *The Sociological Quarterly* 34.4 (1993), 673–693.
Leman, Marc, *Embodied Music Cognition and Mediation Technology* (London: MIT Press, 2008).
Levin, David Michael, *Modernity and the Hegemony of Vision* (London: University of California Press, 1993).
Lowe, Donald M., *The Body in Late-Capitalist USA* (London: Duke University Press, 1995).
Midgeley, Mary, *The Myths We Live By* (Oxford: Routledge Classics, 2011).
Nye, Catrin, 'Torture Evidence Ignored by Home Office, Says Charity,' *BBC News Online*, 21 November 2016. https://www.bbc.co.uk/news/uk-38017463 [Accessed 4 April 2020]
Power, Nina, 'Soft Coercion, The City and The Recorded Female Voice,' *Map Magazine*, May 2018. https://mapmagazine.co.uk/soft-coercion-the-city-and-the-recorded-female-voice [Accessed 22 June 2020]
Sayre, Nathan F. 'The Politics of the Anthropogenic,' *Annual Review of Anthropology* 41 (2012), pp. 57–70.
Schafer, R. Murray, 'The Soundscape,' in *The Sound Studies Reader*, ed. by Jonathan Sterne (London: Routledge, 2012), pp. 95–103.
Thompson, Marie, 'Introduction – Noise: A Useless Concept?' in *Beyond Unwanted Sound: Noise, Affect and Aesthetic Moralism* (London: Bloomsbury, 2017).
Townsend, Mark, 'UK "Ignoring" Systemic Evidence of Torture Among Asylum Seekers,' *The Guardian Online*, 14 March 2010. https://www.theguardian.com/uk/2010/mar/14/asylum-torture-evidence-ignored [Accessed 4 April 2020]

Further Resources Mentioned

https://righttoremain.org.uk
https://righttoremain.org.uk/toolkit/detention/
https://www.compasscollect.com
https://www.freemovement.org.uk
https://www.gov.uk/claim-asylum/after-your-screening
https://www.refugee-action.org.uk

9
SOUNDWASHING

Matt Lewis

Introduction

Loops among acoustic design, Soundscape Studies, and Acoustic Ecology are long established, but as in other areas of design, ideas and practices with roots in environmentalism and the creative arts risk appropriation in order to serve other aims and agendas. The purposeful integration of acoustic features into built environments, such as the ubiquitous acoustic design solution in the form of the water fountain, or piped bird song into liminal spaces such as walkways, resonates with sound art practice, but is I argue, an example of a trend whereby sound practice is being co-opted to fulfill increasingly commercial agendas. These attempts at manipulating our listening experience and the promotion of reductive approaches to our relationship through sound in the world often transplant visual thinking to sonic experience and follow a fashion within design that promotes the designer as a problem solver, guru, and expert, rather than a collaborator. By following a solutionist logic, such modes of operation cannot hope to be inclusive or meaningfully engage with the complex diversity of the affordances of communities and individuals.

Acousticians and sound designers exercise power when it comes to the composition of our listening experience, and current practice elevates designers to a level of expertise that both reinforces inequalities and hierarchies and exposes wider issues concerning design practice. Important work by (Wright, 2022), (Labelle, 2020), (Voegelin, 2014, 2018), (Henriques, 2011) and (Goodman, 2012), all point towards ways we can move away from taxonomical and canonical approaches to sound by embracing the fluid possibilities of listening itself. Following Labelle and others and drawing on principles

DOI: 10.4324/9781003194453-10

from movements such as the Design Justice Network (Design Justice, n.d.), as well as long-established practice in music such as Community Music (Community Music, n.d.) and in theatre such as Cardboard Citizens (Cardboard Citizens, n.d.). I contend that sound has affordance in adding to the emerging critique of design practice. The following is not an argument against more consideration of sound in the construction of our built, work, or home environments, but rather a reading of what the employment of a dominant design logic tells us about the current state of sound in design and beyond.

Commentaries on the contemporary relevance of soundscape studies and acoustic ecology by authors including (Kreutzfeldt, 2010), (Goodman, 2012) and (Kohut, 2015), acknowledge the importance of the work of R. Murray Shafer (Schafer, 1977), and others in the field of acoustic ecology, but problematise hierarchical approaches to listening that don't account for either the complexity of listening experience, or the sociocultural importance of embracing a variety of sounds. Both the prioritisation of certain sounds over others and the idealisation of the natural environment have become key points of fixation in sonic urban design through the integration of acoustic features into the built environment. Added to the post-digital mix, immersive sound media, mobile devices, and digital platforms offer both exciting affordances and the potential for manipulation and normalisation of listening experience. For example, through actions such as public consultations, immersive media, including binaural and ambisonic playback, and arrays are now being used to present speculative listening experiences. In the cases of strategically placed features such as fountains and with public consultations, which create high-resolution, listening experiences using expensive technology, we encounter situations where listening becomes entwined with commercial agendas and sound becomes a solution rather than a binding communicative force.

Soundwashing

The water fountain is a favourite among acoustic solutions in urban planning, with fountains being used to mask "unwanted" sounds and to provide what is often described as "pleasantness" (Lee & Lee, 2020; Miljković, 2019), in an environment. Here, sound is treated as a resource to be used in order to divert attention away from unwanted noise. In the example of the fountain, we come across the familiar visual equivalence in design which regards sound as an object rather than a sensory event, as noted by Voegelin (Voegelin, 2010), experienced through a conglomeration of contingent perceptual responses. Locating sound as a resource also reinforces an extractive logic, which sets the natural world up as a place to be taken from and exploited rather than as part of ourselves and therefore to be lived with in harmony and care. In his poem, *Pleasant Sounds* (Clare, 1820), John Clare paints a bucolic aural portrait of the English countryside, one which reinforces historical and romantic notions of the rural environment. In Pleasant Sounds, everyday onomatopoeia is celebrated, such as the "rustling", "pattering", and "whizzing", sounds of countryside life. The focus on the countryside and the sound of the natural world is one that still prevails today and is exploited in urban design practice through the elevation of natural sounds as an ideal. The fountain, with sonic links to naturally occurring features, such as waterfalls or flowing rivers, the water fountain has a long history in many cultures and is a go-to solution for noise masking in many contexts. Not far from the birthplace of Clare in the UK's City of Peterborough in Cathedral Square, we find bonds between commercial agendas and the use of fountains, here the placement of the fountains is strategic and demonstrates an economic imperative:

> Cathedral Square is very large, so the movement and focus provided by the fountains are space-filling and a place-making attraction. Fountains are expensive to maintain but their impact here is appreciated by all and supported by retailers....After the square was complete and its attractiveness manifest, then retail demand dramatically and demonstrably increased. Major chains have moved in and have added to the variety and liveliness in the square.
>
> *(Academy of Urbanism, 2014)*

Ties between fountains and commercial agendas are also audible in quasi-public and privately owned public spaces; these spaces have the appearance of public spaces, but rites are more regulated and controlled. Exemplifications of such spaces can be found in London in Granary Square, Kings Cross and in White City Living in White City. In Granary Square, the fountains went up early in the new development of the project in 2012, and their placement can be seen as calculated in that they were employed with the affordance of detracting attention from the sounds of the large building development in the area. In White City, in the new development White City Living, the fountains and other water features are at the fulcrum of the new housing development, which revolves around ponds, streams, and fountains. This development was inspired by the 1910 Japanese Exhibition, which at the time also traded on one-dimensional orientalist views about nature. The site is also currently shared with Westfield, one of the UK's largest shopping complexes. As with Peterborough, in both cases, national and global chains surround the fountains, continuing the process of commodification of our shared spaces through soundwashing. Pratt (Pratt, 2017) describes how in these relatively new quasi-public spaces, the shopping centre or mall is now rendered outdoors, integrating retail and strategically placed features to create the deception of a freer, public space. Fountains add to the illusion, and their strategic use implicitly links to a neo-liberal agenda and embodies attempts at standardisation inherent in much design practice examined by Lewis (Lewis, 2020)

> This consumption of acoustic space through the privatisation of aesthetic experience can be seen as an inevitable impact of the neo-liberal program through which, not only are the conglomerate parts of our cultural experience reduced to the lowest common denominator but the perceptual space in which these operate becoming subsumed to form part of the process. Lefebvre, who in The Production of Space (Lefebvre, 1991) extends Marxist concepts around the commodification of leisure to the commodification of space and explains how space in the form of a commodity is standardised and measured. In contemporary listening practices, as exemplified by the fountain, it is this standard format of the water-feature, like Jonathan Sterne's MP3 (Sterne, 2012) or Matt Fullers standardised media object (Fuller, 2005), that represents the measured commodification of our acoustic space.
>
> *(Lewis, 2020, p. 48)*

It is probable that fountains that were historically used primarily for drinking water and washing were also used because of the sounds created by them, but the contemporary employment of fountains for their sonic qualities transforms them into media objects, like those used in personalised listening at home, on public transport, or at work. In Mack Hagood's extensive work (Hagood, 2019), on the use of sound to mask, he explores how environmental recordings and mechanically or synthetically produced white noise are used in a variety of ways, including to aid sleep and mask the effects of conditions such as tinnitus. For Hagood (Hagood, 2019), masking is a form of suppression and points to important power plays concerning listening and design where; "orphic media fight sound with sound to pacify space for beleaguered subjects" (Hagood, 2019, p. 7). In the case of the fountain, as with white noise generators or YouTube videos of the sounds of the sea, this pacification diverts our attention away from important information contained in the sounds we are masking and is an attempt to bind the listener to environments of consumption and employment. Simplistic assumptions about our sonic experience, which reduce sound to a functional object, also discount the contextuality of listening and neglect the affordance of the process, which can potentially exclude listeners. For some, such as people with hearing loss, visual impairment, or neurodivergence, sounds produced by a fountain may create barriers. Features that might be designed to enhance the listening experience of the dominant social group can present dis-affordances for minority groups. As Louise Fryer reflects, "sensory loss leaves you vulnerable" (Lewis 2015a, p. 18), and a normative approach with a lack of inclusion of individuals with a range of disabilities and complex needs perpetuates a singular view of the hearing needs of all listeners. By treating all listeners as the same, as Drever (Drever & Hugill, 2022) terms "auraltypical" or "auditory normates" (Renell, 2022), we negate our understanding of listening as highly contextual and potentially marginalise and alienate. Listening reminds us of the embodied affordance of the process, whereby no environment sounds the same from one moment to the next or is experienced the same way by the same listener from one moment to the next. Birdsong, another, for example, can be stressful and anxiety-provoking for some people, as was explained to me during a field recording workshop with the deafblind charity (Sense, n.d.). In Design Justice (Costanza-Chock, 2020), Sasha Costanza-Chock frames experiences where designed features alienate users as micro-aggressions, which "reproduce the matrix of domination; reaffirm power inequalities" (Costanza-Chock, 2020, p. 44). Lack of an inclusive approach to the integration of sonic elements and singular assumptions over the effects of a design solution that doesn't acknowledge the individuality and contextuality of experience, can thereby add to a whole range of other factors that marginalise individuals or groups on a daily level. The lack of consideration for a diversity of needs within acoustic design mirrors wider issues across design and favours formal, structurally acceptable uses of space. (Manjeya et al., 2021), points out how factors as diverse as street design and refuse collection diss-affords groups and individuals, including street vendors, thereby adding to the exclusion of individuals through the support that favours those holding power. For some time, critiques of the universal approach to spatial design have highlighted the need for an intersectional understanding which accounts for a multiplicity of publics; writing over 20 years ago, Campbell and Marshall highlighted how the idea of a homogenous public hides normative approaches which favour dominant groups:

> As a field, urban design has yet to come to terms with its own complicity that defines the public in universal terms and lacks the capacities to grapple with intersectional injustices. Moreover, a reliance on outmoded conceptions of the public interest helps to conceal power relations that perpetuate social injustice.
>
> *(Campbell and Marshall, 20.2)*

Acoustic considerations also generally only take into consideration an idealised listener, rather than accounting for the complexity of ever-shifting needs, and for Renell (Renell, 2022), the implications of a one-dimensional, or "normative" approach to sound in the built environment, highlights a visual dominance and demonstrates how design can "dictate how and when communications can take place" (Renell, 2022, p. 59). The addition of sonic features, such as fountains, for example, as well as potentially creating barriers to communication, also reconfigure socio-spatial relations, affecting how we feel, operate, and relate to spaces. In the same way that physical features modulate socio-spatial relationships, sounds modulate socio-sonic relations. Added to the complexity here is the temporal-contextuality of listening; no space ever sounds the same from one moment to the next, and due to the constantly shifting elements of a soundscape, what might be an inclusive sonic feature in one moment may add to the sonic web of disaffordances the next. In work with deafblind charity Sense (Lewis, 2015b), what became very apparent was the diversity of responses to listening experiences, which ranged from highly positive and euphoric to scary and anxiety-provoking to the same sounds. These responses also shifted depending on many other factors, sonic, spatial, and otherwise. Therefore, what might be a sonic design solution for one person might create a barrier for another. Sound is a spatial medium, and it can be argued that spatial justice is also sonic justice, with processes that exacerbate spatial injustices and inequality potentially disaffording sonically. The lack of inclusive participation is fundamental to the perpetuation of normative approaches across urban design, and whilst participatory activities might form a part of trends such as design-thinking, even when these activities are employed as part of a design process, the participation may fail to include a diverse group of individuals, or the participation is used to validate an existing view or agenda. Whilst we continue to treat sound as a problem, to be solved by experts, we remain trapped in a design continuum which ultimately favours the needs of developers and businesses whilst using perceived participation as a veil behind which to push through singular purposes. Design deals with perceived problems, many of which are generally indicators of socio-economic conditions, and by detaching indicators from the conditions which produce them or using ornate sticking plaster approaches to issues such as road traffic noise, we do nothing to deal with the underlying issues. Community movements within design, such as the Design Justice Network, offer slower, more considered approaches that ask for design processes to be in the hands of communities, whereby participation is not a tag-on but at the centre of design and therefore responds to socio-economic issues:

> Design justice practitioners flip the "problem" of how to ensure community participation in a design process on its head to ask instead how design can best be used as a tool to amplify, support, and extend existing community-based processes.
>
> *(Costanza-Chock, 2020, p. 84)*

The point here about existing practices is crucial to a reformulation of participatory practice within sound design, towards horizontal ways of working that are led by communities rather than companies or studios acting on contract for developers. Sound-making in the form of music has strong links to other forms of participatory practice, and the kind of practice promoted by networks such as the Design Justice Network echoes approaches with deep roots in art forms such as theatre and music. In many folk music traditions, for example, whilst virtuosity and skill are revered, the hierarchical distinctions between different performers and audiences are more horizontally arranged and can shift from moment to moment. In many folk or jazz sessions, for example, anyone can sit in and play, and the roles of the audience shift both physically and performatively. Community pedagogy in music and theatre is also well established, whereby part of the processes of professionalisation is the sharing of knowledge with less experienced practitioners, thereby building a community of practice. These approaches are innate in sound practice, and I argue that our practice offers particular affordance in rethinking approaches to design, towards approaches to community-led design due to the connective power of sound and the need for an embodied approach to design due to the contingent nature of the listening. Participatory approaches to encountering sound in the urban environment are long established, and activities such as soundwalks offer an opportunity for critical, corporeal engagement with listening in the built environment. However, not only are simplistic readings of the results of these encounters, such as an affinity to the sound of water or birdsong, being extracted, but there also needs to be much more done in order to include a broader range of people in these activities, and also in a broader range of approaches that form these activities. Additionally, as well as wider participation, following Voegelin (Voegelin, 2014; Renell, 2022; and (Drever & Hugill, 2022), this approach needs to be aurally led, and as such, guided not just by the ear but by the body. (Gissen, 2023), in extensive work on disability and design acknowledges the importance of attention to sound in reconceptualising critiques of existing approaches to design, but also asks us to go further:

> Other work cultivates another perceptual sense of form and space via the experience of sound, among many possibilities. I believe that this work, as powerful as it is, simply provides access to the experience, and therefore the concept, of architectural form. By contrast, a disability critique of formalism would extend beyond questions of access *to* form or the sculptural power of accessible forms. A disability critique of formalism would question the values instilled into the concept of form itself as a type of experience.
>
> *(Gissen, 2023, p. 92)*

The critique of formalism offered through a disability critique echoes critiques within sound studies of existing approaches to sound and listening, these include calls for a move away from visual metaphors and perceptual frameworks made by Voegelin (Voegelin, 2014; Henriques, 2011; and (Ingold, 2007). For Ingold the visual metaphor of the soundscape has "outlived its usefulness" (Ingold, 2007, p. 137). For Voegelin, the political power of sound is in opening up questions of inclusivity

> For this "'everyone at all" to truly be everyone at all, it needs to include also the inaudible: that which for physiological and cultural, but also for ideological and sociopolitical reasons we cannot and do not want to hear.
> *(Voegelin, 2018, p. 106)*

For Henriques, an engagement with sound or sounding goes way beyond embodied perceptual responses towards a " sonic logos", which "The *sonic logos* is an attempt to articulate this relationship between the material and mental senses of sound" (Henriques, 2011, p. 243).

An example of the need for a radically different approach to participatory activities was regularly encountered during workshops with deafblind individuals when working with the organisation Sense. Many of these workshops involved field-recording trips, using a variety of sound capture equipment, and we encountered frequent questions about the validity of individuals with hearing loss using headphones and microphones. However, many participants responded extremely positively to recording activities, and these proved the most popular of all. What this experience highlights is not just the importance of not making assumptions but also the importance of inclusivity across the board. It could be easy to understand how reductive modes of practice, such as the installation of fountains, might come about. Another experience from workshops I've facilitated and participated in has been that when discussing favourite sounds, water and birdsong are consistently popular. However, what we also experienced in these workshops and accompanying reflective listening sessions was that, especially when listening to recordings collected during recording walks, many other less obvious sounds were favoured. These included sounds from bicycles, markets, trains, and even air-conditioning, the bane of acousticians' lives. This more inclusive collection of responses, which goes beyond obvious examples, again demonstrates the contextuality of listening and shows that even sounds normally considered unwanted go towards constituting the identity of an environment and that attempts to either mask or replace.

Sound as a Resource

Categorisation of sounds into segregated groups reinforces the separation between us, other animals, and the natural world. By grouping sounds into geophonic, biophonic, and anthrophonic, for example, we don't account for the affordance of the listener to identify the sound source as anything they wish at any given time. Nor do we account for the complexity of sound production, for example, a water fountain may have sonic features of the geophonic world, but the hidden, complex system of piping and sometimes, digitally controlled automation links the fountain to other infrastructures and networks from the anthrophonic and computational worlds. More importantly, the segregation of the built human world from the rest of nature discounts the materiality that makes up our entangled world. We are materially part of so-called geophonic and biophonic worlds, and as Yuk Hui (Hui, 2017) points out, the cultural construction of a single, standardised nature is the product of European, Kantian philosophy and is linked to industrialisation:

> Kant somehow assumes one single nature, which reason compels us to recognize as rational; the rationality corresponds to the organicist teleological universality ostensibly realized in the constitution of both morality and the state. This enchantment of nature is accompanied by a disenchantment of nature, driven by the mechanization enforced by the Industrial Revolution.
>
> *(Hui, 2017, p. 3)*

Listening is powerful in that it affords us connections with the complex entanglements that weave between nature, sound, and media, and as a vibrational continuum (Goodman & Ikoniadou, 2019), sound offers ways of being in the world that move beyond culturally constructed oppositions and delineations towards vibrational connectivity where distinctions between technology, nature and ourselves become redundant. The opposition between ourselves and sound is also audible in the language used in acoustics, terms such as "masking", "mitigation", and "attenuation", to name but a few, all set up sound as something to be contained and controlled, rather than something that is part of us and connects us with the world. The logic contained in this mode of language reinforces a perspective of sound as the other, of something to be contained, to be fearful of, rather than something that connects us.

Using sound as a resource is also consistent with approaches taken by designers in other areas of practice, such as audio branding. Some designers, such as Julian Treasure (Treasure, 2000), promote using the sound of birdsong or classical music as a tool to increase productivity in the workplace. Not only is the notion that music can aid productivity at work problematic concerning personal freedom, but the lack of accountability for a diversity of listening experiences echoes approaches from acoustic design in the built environment explored earlier. The design firm Sensory Experiences recommends playing recordings of the seaside; soft, crashing waves and seagulls, (Sensory Experiences), in seafood restaurants in an attempt at enhancing the customer's eating experience, the seaside becomes reified in order to support the consumption of a high-end eating experience. Again, just as in *Pleasant Sounds* (Clare, 1820), we find the promotion of an idealised, normative, sonic portrait of the seaside. As someone who lives by the seaside can confirm, not only are the sounds of softly crashing waves and seagulls just a small feature of the soundscape.

The explosion and the hype around immersion have seen renewed interest in recording and playback processes such as binaural and ambisonic, which have long been key features of sound practice. Formerly hidden away in music and science departments or in acoustics consultancies, ambisonic and multichannel projects have sprung up, such as Call & Response and Amoneus in the United Kingdom. Binaural audio is also gaining wider popularity thanks to VR and online platforms such as YouTube and Facebook. Auralisation, which simulates the sound of proposed building developments, offers consultancies the opportunity to share and discuss possible future sonic scenarios, and if used inclusively, this technology could provide space for communal listening and engagement around sound issues. However, much of this technology, in particular ambisonic arrays, microphones, and high-quality recording devices are prohibitively expensive, and if speculative auralisation, which engages public stakeholders, is used to sign off large commercial developments, then these listening sessions become problematic. The beauty of sonic composition is the potential to which the composer or artist can re-compose, process, edit, and add sounds, and as cinema has taught us, even the slightest changes in sound design can have a myriad of effects on the audience. Additionally, asking the public to become complicit in the bureaucracy around development, without full engagement in all parts of the processes can become extractive, whereby individuals are given the illusion of choice and participation. The architectural consultancy Arup estimated that 25,000 people listened to their auralisations of the effects of the proposed HS2 high-speed rail development being built in the United Kingdom to link London and the North-West (Arup, n.d.). One significant issue with such approaches is that choices for listeners are predominantly based on recordings made by presumed experts. This approach treats listening as passive, where an expert with access to expensive and exclusive technology creates a theatrical experience that listeners are expected to accept without sufficient information about the underlying compositional processes. Hagood (2019) aligns the technologically determined approaches with the neoliberal programme, "the very presence of a technological" solution

> To this problem of conflicting freedoms reinforces the essential neoliberal belief that problems must be solved individually and within the market rather than addressed as systemic issues: individual consumption, rather than collective action, is the site of social agency.
>
> *(Hagood, 2019, p. 189)*

Tech fetishism and a focus on tools rather than communal processes, by necessity, isolate communities and individuals from choices around design and its implications for communities, as it emphasizes the technology employed rather than communal knowledge. For Hui, solutionist approaches that focus on tools created to solve problems follow a logic that can be critiqued by other philosophical systems, such as Daoism. In writing about relationships between contemporary technology and modes of thought in China, Hui finds a critique of technologically determined approaches through Daoist philosophy. In his enquiry, Hui aligns Dao with Way or Method and Qi with Tool, he explains how in Daoism "Dao is more fundamental than Qi (the tool)" (Hui, 2017, p. 2). For Hui, therefore, the technological tools must follow the appropriate way of working, otherwise imbalance occurs.

Sonic Spatial Justice a way forward

Within current dominant socioeconomic and sociopolitical paradigms where design is so tightly entwined with narrow commercial aims, we may not see a seismic shift in sound design in the near future. However, it is still important to keep asking what listening tells us about the designed world, and perhaps what sound practice reminds us, is that instead of building platforms, interfaces, solutions, and other resources, the most fundamental is coming together to listen, make noise, and critically problematise what our ears tell us about the world that is being designed for us. Community participation within design does not necessarily equate to equality and diversity of outcomes. With participatory activities now common components of trends such as design thinking, we need to ask how activities and praxis with century-old roots in the creative cultures of theatre and music can resist co-option by processes that reinforce sonic inequality through the tokenistic use of participatory practice. The Brazilian group Design & Oppression Network [Design e Opressao, n.d.], which was founded in 2020, offers not only an important reminder of the need to resist inequalities within design practice but also the need to critique the appropriation of knowledge and practice itself. Drawing on the rich genesis of Latin-American and Brazilian community pedagogical practice, including the work of Freire (1972) and Boal (2005), the network examines ways in which the practice and methodologies of thinkers and practitioners such as Freire and Boal are being appropriated by design thinking. This important work reminds us that designers, therefore, need to honour not only the needs of communities but also the local and global community of knowledge around participatory practice, in order to perpetuate the extractivist nature of commercial design. Much of the work of the Design & Oppression Network utilises online communication platforms to connect and build a community to embrace the possibilities of "designing for liberation", in doing so, they offer not only solidarity and support but also the opportunity of free online training to support the growth of the network and the empowerment of individuals outside of academia. Pervasive platforms and devices can support connection across continents and offer some very useful options in terms of accessibility; however, faster, wider communication and distribution of tools don't necessarily remove barriers and can reinforce power equalities (Alper, 2017), and (Lewis, 2020). Despite this, there is meaningful potential in building communities of sound through digital platforms and on mobile devices; SomaRumor (SomaRumor), another Latin-American initiative, uses a variety of platforms and actions to build a participatory community that engages critically through sound. In 2021, responding to the global pandemic, SomaRumor initiated a free pedagogical initiative that brought practitioners together from across Latin America and beyond. One of the powers of this project was the utilisation of video conferencing to provide spaces to listen and be heard in addition to activities such as networked music-making and critical listening sessions. In the United States and beyond P5.js (P5.js, n.d.) and Processing Foundation (Processing Foundation, n.d.) have been building an open-access global community around coding since 2001 and with tone.js are joining the FLOSS (Free/Libre/Open Source Software) sound community, which already has strong networks built around tools including TidayCycles (Tidal Cycles, n.d.), SuperCollider (SuperCollider, n.d.), and Pure Data (Pure Data, n.d.). Although often built around platforms or design tools, these examples represent how slower, horizontal approaches can differ from the top-down approach, whereby expensive tools are employed. As demonstrated through Deep Listening, Oliveros (Oliveros, 2005), however, sonic immersion is not contingent on perfectly constructed speaker arrays; immersion is everyday, and group listening, sound walks, field recording, and musical improvisation can all play a part in the making of our worlds. By dismantling the barriers between experts and non-experts in sound design through collaborative sound collection and composition, these activities indicate the potential for a much more nuanced, complex, and integrated participatory approach to sonic design. The affordance of working with sound to critique design is that it allows us to move away from semantic language and begin to build sonic worlds together with others. Sound offers a means to move away from forms of standardisation in formalism and semantic language and the typical statistical approaches commonly employed in sound and noise consultations, and instead immersing ourselves in the materiality of sound through recording, composition, and musical improvisation, the resulting soundscapes became much more inclusive, complex, and diverse.

References

Academy of Urbanism. (2014). *Cathedral Square|Peterborough|The Academy of Urbanism*. Available at: https://www.academyofurbanism.org.uk/cathedral-square/ [Accessed 23 June 2023].

Alper, M. (2017) *Giving Voice: Mobile Communication, Disability, and Inequality*. Cambridge, MA: The MIT Press.

Arup. (n.d.) *SoundLab*. Available at: https://www.arup.com/services/digital-solutions/soundlab/ (Accessed: 21 July 2025).

Boal, A. (2005). *Games for Actors and Non-Actors*. London: Routledge.

Cardboard Citizens. (n.d.). Cardboard Citizens. Available at: https://cardboardcitizens.org.uk/ (Accessed: 15 June 2023)

Clare, J. (1820). Pleasant Sounds. In *I Am: The Selected Poetry of John Clare*. Farrar, Straus and Giroux, p. 291.

Community Music. (n.d.). *Community Music* [online]. Available at: https://www.cmsounds.com/ (Accessed: 21 June 2023)

Costanza-Chock, S. (2020). *Design Justice, Community-Led Practices to Build the Worlds We Need*. Cambridge, MA: MIT Press.

Design e Opressão. (n.d.). *Design e Opressão*. Available at: https://www.designeopressao.org/ (Accessed: 21 June 2023).

Design Justice Network. (n.d.). *Design Justice*. Available at: https://designjustice.org/ (Accessed: 21 June 2023).

Drever, J.L. and Hugill, A. (eds.) (2022). *Aural Diversity*. Abingdon: Routledge.

Freire, P. (1972). *Pedagogy of the Oppressed*. New York: Bloomsbury Academic.

Gissen, D. (2023). *The Architecture of Disability: Buildings, Cities, and Landscapes beyond Access*. Minneapolis, MN: U of Minnesota Press.

Goodman, S. (2012). *Sonic Warfare: sound, Affect, and the Ecology of Fear*. Cambridge, MA; London: Mit Press.

Goodman, S. Toheys and Ikoniadou, E. (2019). *Unsound: undead*. Falmouth, Urbanomic; Cambridge, MA: The MIT Press.

Hagood, M. (2019). *Hush. Media and sonic self-Control*. Durham: Duke University Press.

Henriques, J. (2011). *Sonic Bodies: Reggae Sound Systems, Performance Techniques, and Ways of Knowing*. New York: Continuum.

Hui, Y (2017) Cosmotechnics as Cosmopolitics. *e-flux Journal*. Available at: https://www.e-flux.com/journal/86/161887/cosmotechnics-as-cosmopolitics/

Ingold, T. (2007) Against soundscape. In: Carlyle, A. (ed.) *Autumn Leaves: Sound and the Environment in Artistic Practice*. Paris: Double Entendre, pp. 10–13.

Ingold, T. (2011) *Being Alive: Essays on Movement, Knowledge and Description*. London: Routledge.

Kohut, T. (2015). Noise Pollution and the Eco-Politics of Sound: Toxicity, Nature and Culture in the Contemporary Soundscape. *Leonardo Music Journal*, 25, pp. 5–8. https://doi.org/10.1162/lmj_a_00924

Kreutzfeldt, J. (2010). Acoustic Territoriality and the Politics of Urban Noise. *Soundscape*, 10, no. 1, pp. 14–17.

LaBelle, B. (2020). *Acoustic Justice: Listening, Performativity, and the Work of Reorientation*. New York: Bloomsbury Academic.

Lee, H.M. and Lee, H.P. (2020). Noise Masking in High Population Country Using Sound of Water Fountain. *Applied Acoustics*, 162, p. 107206. https://doi.org/10.1016/j.apacoust.2020.107206

Lefebvre, H. (1991) *The Production of Space*. Translated by D. Nicholson-Smith. Oxford: Blackwell

Lewis, M. (2015a). *No Such Thing as Empty Space Sound Project Report*. London: Sense.
Lewis, M. (2015b). No Such Thing as Empty Space. Available at: www.mattlewis.info/no-such-
Lewis, M. (2020). How Can We Listen with Our Future? In: *DRHA 2019 Conference Proceedings. Radical Immersions*. DRHA & University of Greenwich.
Manjeya, E., Chirisa, I. and Chavunduka, C.M. (2021) 'Spatial justice and the design of future cities in the developing world', in *The Palgrave Encyclopedia of Urban and Regional Futures*. Cham: Palgrave Macmillan, pp. 1–11
Miljković, D. (2019) 'Evaluation of soundscape perception and pleasantness in urban environments', *Applied Acoustics*, 152, pp. 1–9.
Oliveros, P. (2005). *Deep Listening: A Composer's Sound Practice*. New York: Universe.
p5.js. (n.d.). *p5.js*. Available at: https://p5js.org/ (Accessed: 22 June 2023)
Pratt, A. (2017). The Rise of the Quasi-Public Space and its Consequences for Cities and Culture. *Palgrave Communications*, 3(1), pp. 1–4. https://doi.org/10.1057/s41599-017-0048-6
Processing Foundation. (n.d.). *Processing Foundation*. Available at: https://processingfoundation.org/ (Accessed: 23 June 2023).
Pure Data. (n.d.). *puredata.info*. Available at: https://puredata.info/ (Accessed: 11 September 2023).
Renell, W. (2022). The Auditory Normate Engaging Critically with Sound, Social Inclusion and Design. In Drever, J.L. and Hugill, A. (eds.) *Aural Diversity*. Abingdon: Routledge, pp. 83–94.
Schafer, R. (1977). *Murray: The Tuning of the World*. New York: Knopff. Schafer, R. Murray (ed.): Vancouver Soundscape.
Sensory Experiences UK. (n.d.). *Sensory Experiences UK*. Available at: https://sensoryexperiences.co.uk/ (Accessed: 21 July 2023).
Sterne, J. (2012). *The meaning of a format MP3*. Durham: Duke University Press.
SuperCollider. (n.d.). *SuperCollider Programming Language and IDE*. Available at: https://supercollider.github.io/ (Accessed: 20 June 2025).
TidalCycles. (n.d.). *TidalCycles — live Coding Patterns*. Available at: https://tidalcycles.org/ (Accessed: 1 July 2023).
Treasure, J. (2000). *Sound Business*. Cirencester: Management Books Ltd.
Voegelin, S. (2010) *Listening to Noise and Silence: Towards a Philosophy of Sound Art*. New York: Continuum.
Voegelin, S. (2014). *Sonic Possible Worlds: Hearing the Continuum of Sound*. New York: Bloomsbury Academic.
Voegelin, S. (2018). *The Political Possibility of Sound: Fragments of Listening*. New York: Bloomsbury Academic.
Wright, M.P. (2022). *Listening After Nature: Field Recording, Ecology, Critical Practice*. New York: Bloomsbury Academic.

INDEX

For Product Safety Concerns and Information please contact our EU representative GPSR@taylorandfrancis.com Taylor & Francis Verlag GmbH, Kaufingerstraße 24, 80331 München, Germany